Excursions in Mathematics

Excursions in Mathematics

C. STANLEY OGILVY

PROFESSOR EMERITUS OF MATHEMATICS
HAMILTON COLLEGE, CLINTON, NEW YORK

LIBRARY
ACC No. DEPT.
01035734
CLASS No.
510 OGI
UNIVERSITY
COLLEGE CHESTER
-1023482

DOVER PUBLICATIONS, INC.
New York

Copyright

Copyright © 1956 by Oxford University Press.
Copyright © renewed 1984 by C. Stanley Ogilvy.
All rights reserved under Pan American and International Copyright Conventions.

Published in Canada by General Publishing Company, Ltd., 30 Lesmill Road, Don Mills, Toronto, Ontario.
Published in the United Kingdom by Constable and Company, Ltd., 3 The Lanchesters, 162–164 Fulham Palace Road, London W6 9ER.

Bibliographical Note

This Dover edition, first published in 1994, is an unabridged, slightly corrected republication of the work first published by Oxford University Press, New York, 1956, under the title *Through the Mathescope*. The author has added a few footnotes to the present edition.

Library of Congress Cataloging-in-Publication Data

Ogilvy, C. Stanley (Charles Stanley), 1913–
[Through the mathescope]
Excursions in mathematics / C. Stanley Ogilvy.
p. cm.
Originally published: Through the mathescope. New York : Oxford University Press, 1956.
Includes bibliographical references and index.
ISBN 0-486-28283-X (pbk.)
1. Mathematics—Popular works. I. Title.
QA93.036 1994
510—dc20 94-24696
CIP

Manufactured in the United States of America
Dover Publications, Inc., 31 East 2nd Street, Mineola, N.Y. 11501

*The Mathescope**

ASTRONOMERS USE TELESCOPES and laboratory scientists use microscopes to aid them in seeing more clearly the subjects of their research. For the same purpose mathematicians sometimes use mathescopes. But a mathescope is not a physical instrument. It is an intellectual instrument, with reason for its pedestal and inspiration for its lenses. No one has ever seen a mathescope. But then, neither has anyone ever seen an integral, or a geometric point, or, for that matter, a number. You can see the symbol that someone writes down to stand for a number; but a number itself has no earthly physical being to see, touch, or smell. Mathematicians, as we shall see, deal not with tangibles but with ideas. Since the very stuff that mathematics is made of exists only in

* This book was originally published under the title *Through the Mathescope*. [1994 note.]

men's minds, it is fitting that one should view it through a mental mathescope.

The phrase 'Through the Mathescope' was suggested by the distinguished mathematician, Professor (Emeritus) Edward Kasner of Columbia University. It was invented by Watson Davis, editor of the weekly magazine *Science News Letter*, who has kindly consented to its use as the title of this book.

Contents

Excursions in Mathematics

1 / *What Do Mathematicians Do?*

WHAT IS MATHEMATICS? What goes on in that extraordinary world of rules and symbols? We are repeatedly told that mathematics is becoming daily more important as science proceeds on its inexorable way. Yet very few intelligent people have any actual knowledge of mathematics beyond what they gleaned years ago in school, and most of that has been forgotten. The result is a confused jumble of general impressions about the subject, none of which would stand a critical inspection.

A good many people have the idea that mathematics deals with numbers, and that therefore higher mathematics must deal with great big numbers. Their picture of the mathematician is that of a little man who sits all day making enormous calculations in arithmetic. Such a concept could not be farther from the truth.

Even the basic premise is wrong: mathematics does not necessarily deal with numbers. Many chapters—whole books—of highly complex mathematics contain not a single number larger than 2 or 3. Mathematics talks in symbols more often than not, and the symbols frequently do not even represent numbers. It is true that there is one branch of the subject that does deal exclusively in numbers, appropriately called Number Theory. It studies the properties of the whole numbers themselves, and is often forced to tackle very large ones. You will meet a few of them in Chapter 2. But except in Number Theory, you would be surprised to know how few and how simple are the numbers which mathematicians generally handle.

A second popular misconception is that mathematics is a dead science—all worked out. This is probably due to the fact that most people terminate their study of the subject after they have glanced at what was completed by the Arabic, Egyptian, and Greek mathematicians of antiquity. Mathematics in those days was not yet out of diapers. To claim an acquaintance with real mathematics because you have gone through algebra and plane geometry in school is like saying that you are a qualified musician if you can play 'Chopsticks' on the piano.

The mathematician who is rash enough to mutter something to a non-scientific friend about new work that he is doing always gets the same reaction: 'My goodness, how can you possibly do anything original in mathematics? Surely all the theorems must have been proved long ago!' The mathematician now has two courses open to him. He can suddenly remember an important engagement elsewhere, or he can settle down for a long explanation if he thinks his friend is sufficiently interested to listen. *Rigor mortis* has not yet set in on the body mathematical. The fact is that it was never so much alive as it is today; and the indications are that it is still in the early stages of its growth. A statistic or two might not be out of place.

The mathematics department of the Columbia University library receives about 350 journals and periodicals devoted to the subject, written in many languages. Most of these magazines *do not accept for publication* anything except new, original work in mathematics. And some of them have a two-year backlog of accepted papers, with more pouring in to the editors all the time.

Here is another thought for those who have the idea that all the mathematics there is to do has already been done. The most comprehensive existing history of mathematics is Moritz Cantor's four-volume *Geschichte*, which is an outline of the development of the subject totaling some 3600 pages. But it *ends* with the year 1799. It has been estimated that about twenty additional volumes of the same size—five times as much as all the previous history put together—would be required to cover, in similar outline form only, the new work done in the nineteenth century. As for the twentieth century—the same only more so. So enormous is the field today that no one person can claim to be familiar with all of it, or expect to become an expert in more than a small corner of it. Henri Poincaré (1854–1912) was the last great mathematician to work in practically all of the major branches of the subject. Probably never again will anyone be able to make all of mathematics his own, unless great strides are taken in the direction of synthesis and organization of the tremendous system that has been created. We shall have more to say about such a synthesis.

Mathematics is not, then, merely a bunch of calculations that could be done much faster and better by an electronic machine; nor is it a science developed in ancient times, with nothing added to it since then except a bit of polish here and there. A third appellation which it sometimes receives is that of 'a tool for the physical scientists and engineers.' With this view we must have some sympathy; it is at least partly justifiable. In fact it is the truth but not the whole truth. Mathematics is a very necessary tool for the sciences. It has been called the handmaiden of the sciences and without its help they would have had to give up long ago. A large body of 'higher' mathematics—not just arithmetic—is in useful service. The invention of much of it was stimulated by an urgent need for it in the laboratory or the machine shop.

But this is not the whole story. Applied mathematics is only one facet of the great gem. And although the applied scientist has no use for any other kind of mathematics, the pure mathematician on the other hand may have no use for applied mathematics. This is not to imply that there is a feud going on between the pure and the applied mathematicians; but nonetheless there are definitely the two points of view. Who is this 'pure mathematician'? What does he do, and why?

The pure mathematician works at mathematics for its own sake. Sometimes he does not care two hoots whether his work has any outside application whatsoever. Sometimes, if it's applicable to some other field, that's all right too, and he is pleased to have made a 'useful' contribution. But that is not *why* he works at it. It is often honestly difficult for him to explain his own motives. He does mathematics for the same reasons that the artist paints: for the fascination of the subject itself; for the satisfaction of producing something that to himself and to his colleagues is beautiful; and, if he has a spark of real genius, because he can't help it. Mathematics, for all its scientific dress, is very like an art. Only those with a certain special talent are in a position to make permanent contributions, just as only those with special talents can carve statues or write poetry.

If mathematics is a law unto itself, as seems to be suggested, can a creative mathematician strike out on his own and devise some new kind of symbols, on which he does new operations and gets new answers? Yes. And the point is that neither the symbols nor the 'answers' will necessarily have the slightest resemblance or application to anything that anyone has ever seen in 'real life.' Such abstract mathematics is going on all the time. Perhaps this may give you an idea of what is meant by pure mathematics, and why the field is so unlimited.

To do *meaningful* work of an original nature, however, is not quite so easy as we have perhaps made it sound. Even though a new development may not appear to have any possible practical application outside of mathematics, it must, to have any value, have at least some connection with something else which is going on or has gone on within the field itself. Otherwise it would be too much like the modern painting that was so original that nobody knew what it meant; they asked the artist, and he didn't know either. A new theorem that sheds light on an old problem, or provides a connecting link between two other theorems or subjects hitherto thought isolated, or leads the way to further developments—this is the kind of work that is hailed by one's fellow mathematicians and labeled important. The working mathematician is always on the watch for announcements of such results. They may provide him with just the required machinery, which he had sought but had been unable to find, to break down a nasty barrier that stands in his path, so that he can now pull an old paper out of the bottom drawer, dust it off, and complete a fine bit of theory which he had been forced to abandon five years before.

But let us get back to one more, and perhaps the greatest, popular misconception about mathematics. This is that it can be checked in the laboratory. This is the fallacy that held back the development of geometry for 2000 years. Everyone—mathematicians included—thought that Euclidean geometry, which is the ordinary plane and solid geometry that you learned (or perhaps I should say 'were exposed to') in high school, was *the* geometry 'because it fits the universe.' This is nonsense. For one thing, it happens not to fit the universe, as Einstein and others have shown. But, far more important, whether it fits or not has nothing whatever to do with how 'right' it is. Euclidean geometry, when made properly rigorous under modern treatment, is a very fine geometry indeed—a perfect mathematical system. But it is perfect because it is consistent within itself, and not because it describes anything outside of itself. *There are other equally sound geometries*. It was this great discovery, made at about the same time by two or three different mathematicians of the last century, that liberated geometry from the shackles that had bound it for so long.

Einstein put it neatly: 'So far as the theorems of mathematics are about reality, they are not certain; so far as they are certain, they are not about reality.' How contrary to the layman's point of view! You can test mathematics against itself for inconsistencies within itself; you cannot test it against the physical world. If you try, you are seeking to match two different kinds of things: mathematical symbols and laboratory experiments. That they can be matched so successfully so often is one of the greatest wonders of all. But many times the match is very difficult to find: this is the problem of theoretical physics, which aims to describe the phenomena of nature by means of mathematical equations. The theoretical physicist is farther away from the ultimate solution of his problem than he thought he was fifty years ago. The universe seems not to be the complicated but systematic mechanical gadget that it was formerly believed to be. Listen to Bertrand Russell, in a typical Russellian blast: 'Academic philosophers . . . have believed that the world is a unity. . . . The most fundamental of my intellectual beliefs is that this is rubbish. I think the universe is all spots and jumps, without unity, without continuity, without coherence or orderliness. . . . Indeed, there is little but prejudice and habit to be said for the view that there is a world at all.' And the title of the work from which this is taken is *God Is Not a Mathematician.*

'So much for physics,' says the pure mathematician. Back to his ivory tower he goes, to do some more mathematics. But as Lewis Carroll (a mathematician) said, in a very different connection, 'curiouser and curiouser.' What happens? He turns out a piece of work which is elegant and concise, but which seems to have no conceivable bearing on anything that could happen elsewhere in mathematics, to say nothing of the physical world. Even his colleagues say, 'Pretty. Now put it away and start something different.' So he does. But twenty-five or fifty years later this very work is precisely what someone else needs in the course of a bigger and more 'useful' project. It is a rash prophet who condemns any piece of mathematics as useless. A classic example is the application of Riemannian geometry to relativity theory. During the 1850s Riemann invented his geometry of manifolds and extended the definition of curvature to more than two dimensions. These concepts were then considered by many informed scientists to be nothing but mathematical curios, but had they not been ready to hand in 1905, Einstein would have had to develop them before being able to proceed with his relativity theory. Again, relativity itself seemed a strange and unacceptable interpretation of the physical world during its early years. Among other things, it tied matter and energy together through the famous equation, $E = mc^2$. Matter and energy had long been assumed to be inviolate; the whole structure of physics depended on their not being convertible the one into the other. But convertible they are, as we all know only too well today, the energy of an atomic bomb being obtained by the destruction of some of its matter. The possibility of atomic fission hinged on the validity of the relativity equations, and the great atomic reactors of today follow the dictates of $E = mc^2$.

Even more striking examples of connections between apparently remote topics can be found within the field of mathematics itself. We have mentioned the enormous growth of the subject in the past century or two, resulting in a diversification and specialization of alarming proportions. The greatest living mathematicians are constantly on the warpath against such partitioning, seeking instead some means of synthesizing and unifying. Occasionally a discovery of this nature is made, always with far-reaching and satisfactory results. The development of the theory of groups, for instance, has brought together astonishingly diverse branches of geometry, algebra, and analysis, and permitted them to be studied from a single point of view.

We cannot, in a book intended for the intelligent layman, go into a description of something like group theory. What we can do is to show you a few of the interesting and perhaps amusing aspects of mathematics which lie near enough to the surface so that we do not have to dig very deep to reach them. Whenever possible, we shall point out how some of the apparently disconnected topics are interrelated. We shall not always be able to demonstrate an interconnection, sometimes because the demonstration would be too difficult or the connection too devious, but equally often because no connection is yet known. It is probably the secret hope of most mathematicians that all the myriad topics of the subject can somehow be interwoven, and that some day, if we are smart enough to see it, the whole structure may emerge as one magnificent unit. If this is so, that day is a long way off. But to weave two hitherto unrelated pieces of the fabric together is one of the greatest thrills the study of the subject can provide. If we can show you a few of the cross-connections, and hint at some of the others, our project will have been a success.

A word of caution is in order here. Please do not, after reading this book, carry away the impression that you have been looking over the shoulder of a modern working mathematician. Not much of our discussion will attain the level of 'higher mathematics,' which *begins* (not ends) with the calculus. We shall do our best, however, to indicate from time to time the kind of thing with which a mathematician might concern himself. We hope that before we have finished you may be in a better position to answer for yourself the question posed by the chapter heading, 'What do mathematicians do?'

Mathematics can be roughly divided into four main branches: number theory, algebra, geometry, and analysis. Without trying to define each, we shall let these divisions organize our chapters in a general way, and we shall take a look at each division. In addition, we shall occasionally be able to forge connecting links between the chapters, and even, *mirabile dictu*, between the divisions.

Let us be on our way.

2 | *Advanced Arithmetic*

THE NUMBER 142857 puts on a rather special performance. If it is multiplied by any digit less than 7, the result is a *cyclic permutation* of the original number; that is, the order of the digits remains unchanged, only we begin with a different one. What is meant by the six cyclic permutations of six objects is most clearly illustrated by writing them down:

$$1 \times 142857 = 142857$$
$$2 \times 142857 = 285714$$
$$3 \times 142857 = 428571$$
$$4 \times 142857 = 571428$$
$$5 \times 142857 = 714285$$
$$6 \times 142857 = 857142$$

Note also for future reference that

$$7 \times 142857 = 999999$$

Mathematicians are forever generalizing. No mathematician studying 142857 for the first time could fail to ask, are there any other numbers with this strange property, and if so, how do I find them? In order to answer this question we must peer into the dense and often impenetrable jungle of Number Theory, a seemingly innocent but actually very difficult region of mathematics.

• • •

What is meant by the square root of a number N? It is that quantity which, when 'squared' (multiplied by itself), will give N. For instance, the square root of 25 is 5, written

$$\sqrt{25} = 5.$$

It is perfectly true that -5 is also a square root of 25; but for the present let us talk about positive square roots only.

It is easy to see that $\sqrt{9} = 3$, $\sqrt{4} = 2$, and $\sqrt{1} = 1$. But what about all the numbers in between? What, for instance, about the square root of 2? We seek some number, x, such that

$$\sqrt{2} = x.$$

That is,

$$2 = x^2.$$

Now suppose the answer, x, is a common fraction. Then we can reduce this fraction to lowest terms. Let p/q represent x reduced to lowest terms. These lowest terms exist if any fraction exists; and we must be certain to use them for p and q. You may be willing enough to grant this now; but if I do not insist on it, you will want to hedge later.

Very well: we are looking for p/q such that

$$2 = \left(\frac{p}{q}\right)^2$$

$$2 = \frac{p^2}{q^2}$$

or,

$$2q^2 = p^2.$$

This seems a modest requirement. We seek two whole numbers, p and q, such that the square of one is just double the square of the other. The collection of perfect squares is infinite. The squares begin

1, 4, 9, 16, 25, 36, 49, 64, 81, 100, 121 ···

and go on forever. 49 is almost twice 25. Surely there must be *one, somewhere,* that is exactly twice another one? Strangely enough, there isn't.

The proof is not difficult. It has been known since Greek times.* Notice that, in the last equation, the left hand side is an even number (all multiples of 2 are even). Hence the right hand side, p^2, is an even number. A little thought will convince you that therefore p has to be an even number. (A proof of this is given in the Notes; but it is almost obvious that the square root of an even number cannot be odd.) If p is an even number, it is itself a multiple of 2. Therefore p^2, since it contains 2 twice as a factor, is a multiple of 4, say $4k$. We can now write

$$2q^2 = 4k$$
$$q^2 = 2k.$$

But now this says that q^2 is even, and hence q is even. Thus we have shown that both p and q are even numbers: they both contain the factor 2. This says that the fraction p/q, which I insisted

* Generally credited to Euclid, *c.*300 B.C., although it was known to the Pythagoreans.

a moment ago had to be in lowest terms, is after all not in lowest terms.

What does such a contradiction mean? It means that we made a series of reasonable statements and deductions *based on the original assumption* that $\sqrt{2}$ is some common fraction. It turned out that we arrived at a result that contradicted one of our own statements. Nothing else was wrong; so the only other possible conclusion is that the *original assumption* was wrong, and $\sqrt{2}$ is *not* expressible as a common fraction.

This is what is meant by the statement 'the square root of two is irrational.' An irrational number is not the quotient of any two whole numbers.

• • •

Mathematicians are often asked, 'How can you possibly prove that something can't be done? The fact that *you* haven't been able to do it is no proof!' Of course it isn't. To say that something has not been done is entirely different from saying that it cannot ever be done. It is not impossible to travel to Mars, although the through service is not yet in operation. It *is* impossible for a plane triangle, let us say, to have its three sides of lengths 2″, 3″, and 50″ respectively.

We have spelled out the proof of the irrationality of the square root of 2 because it is a nice example of a proof of mathematical impossibility. The reader who understands the foregoing section is not likely to spend many hours looking through a table of squares in a search for one that is double another. Yet this is exactly the futility level of, for instance, angle trisectors. Anyone with sufficient mathematical background can understand the proof that, with ruler and compass as the only tools, a random given angle can*not* be trisected. But this does not seem to deter the would-be trisectors. We shall have more to say about trisections in Chapter 10.

A famous English mathematician of the last century, Augustus de Morgan, once delivered a lecture on this subject, after which a gentleman of the audience was heard to remark, 'Only prove to me that it is impossible, and I will set about it this very evening.' De Morgan's style was as lucid as it was amusing. His listener must have been singularly thick-headed to have missed the point so completely.

• • •

The Pythagoreans (*c.*500 B.C.) were greatly disturbed by the existence of irrational numbers. They knew about $\sqrt{2}$; it stared them in the face whenever they drew the diagonal of a unit square. Here was clearly a line of length $\sqrt{2}$, because by the famous right triangle theorem that bears the name of the Pythagorean school, its length had to be $\sqrt{1^2 + 1^2}$. (Fig. 1.) But in Greek times (and for many centuries thereafter) there was something mystical and sacred about numbers. Rational numbers (common fractions) had a 'rightness' about them not shared by the irrationals. Furthermore, the Pythagorean theory of magnitudes was based on *commensurable* ratios, and we proved just now that $\sqrt{2}$ is not commensurable with any integer.

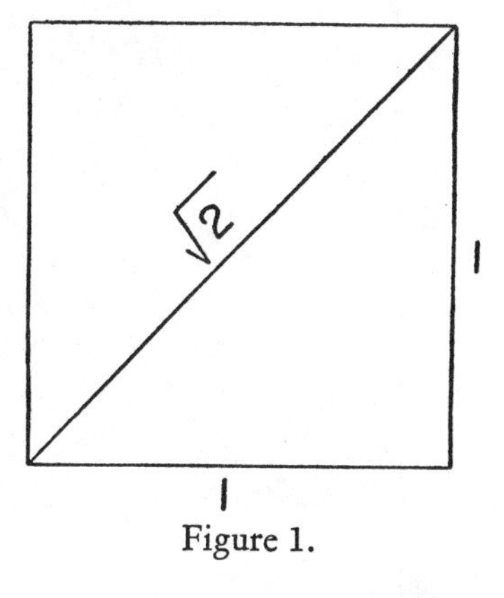

Figure 1.

One story has it that the Pythagoreans suppressed their knowledge of the irrationality of $\sqrt{2}$, and went to the length of killing one of their own colleagues for having committed the sin of letting the nasty information reach an outsider. If this is true, it is a most interesting commentary on the intellectual (dis)honesty of the times. Today any discovery that upsets previously held notions is widely hailed and thoroughly investigated, not suppressed. Mathematicians have learned by experience that when a theory has to be revised because a point of weakness has been detected, the structure that rises on the strengthened foundation is likely to be more beautiful and useful than the old.

• • •

The following exercise demonstrates the geometric meaning of the statement that the diagonal of a square and its side are incommensurable quantities.

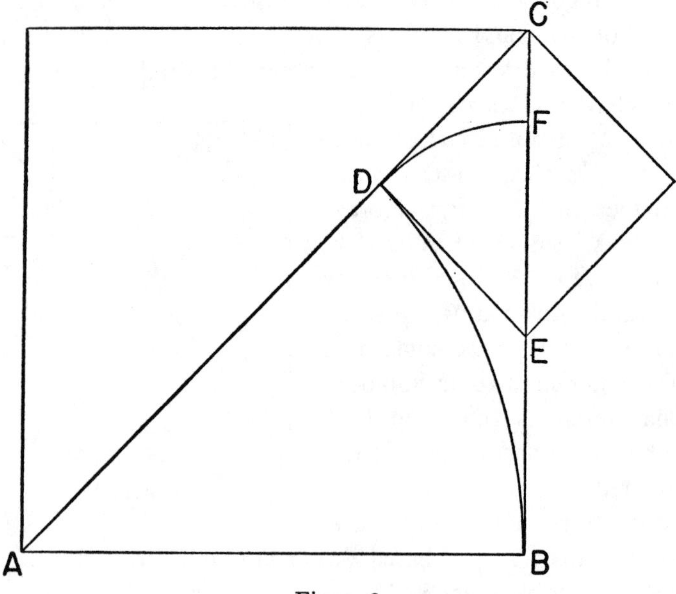

Figure 2.

The diagonal of a unit square is $\sqrt{2}$ (Fig. 1). But all squares are the same shape (similar). Hence we can say, following the lettering of Figure 2,

$$\sqrt{2} = \frac{\sqrt{2}}{1}$$

$$= \frac{AC}{AB}$$

$$= \frac{AD + DC}{AB}$$

$$= \frac{AD}{AB} + \frac{DC}{AB}$$

$$= 1 + \frac{DC}{AB}$$

$$= 1 + \cfrac{1}{\cfrac{AB}{DC}}$$

$$= 1 + \cfrac{1}{\cfrac{BC}{DC}} \qquad (1)$$

Now: $DC = DE$ (sides of a square),
and $DE = EB$ (tangents to a circle from an external point).
Also $DE < CE$ (read 'DE is less than CE').

Hence $BE + EF < BC$, which says that DC divides into BC twice ($BE + EF$) with remainder CF. That is, in the language of division,

$$\frac{BC}{DC} = 2 + \frac{FC}{DC}$$

Hence, from the line marked (1),

$$\sqrt{2} = 1 + \cfrac{1}{2 + \cfrac{FC}{DC}}$$

$$= 1 + \cfrac{1}{2 + \cfrac{1}{\cfrac{DC}{FC}}} \qquad (2)$$

But now, by proportional parts of similar figures (similar *triangles* if you wish to complete the diagram),

$$\frac{DC}{FC} = \frac{BC}{DC}$$

That is, the last fraction of (1) is the same as the last fraction of (2). Hence if we continue the process we will never get anything different. We can express this idea by writing

$$\sqrt{2} = 1 + \cfrac{1}{2 + \cfrac{1}{2 + \cfrac{1}{2 + \cfrac{1}{2 + \cdots}}}}$$

where the three dots, as usual, mean 'and so on, forever.' This expression is known as a non-terminating continued fraction.

A little cogitation should make it plain that what we are really seeking by this process is a common measure (greatest common divisor, if you prefer), of AB and AC; and that none exists if the process does not terminate. If the continued fraction did terminate, it could be reduced backward, by simple arithmetic, to a single common fraction (rational number). Although we have not proved the converse, it happens to be true: a non-terminating continued fraction represents an irrational number.*

If you are feeling eager, develop by this method the geometry of the square root of three. Prove geometrically that

$$\sqrt{3} = 1 + \cfrac{1}{1 + \cfrac{1}{2 + \cfrac{1}{1 + \cfrac{1}{2 + \cdots}}}}$$

* For an easily understood proof, see Courant and Robbins, *What Is Mathematics?* (Oxford University Press, New York, 1948), pp. 49, 301. This book is a masterly presentation of a considerable mass of real mathematics in a form accessible to the reader with a modest technical background.

What is meant by the sum of an *infinite* series? Surely if you keep on adding positive quantities, the result gets bigger and bigger without limit. Obviously. But the trouble is, in mathematics, so many of the things that are 'obviously' true aren't true.

Take a line one foot long (Fig. 3). Bisect it. Bisect the right

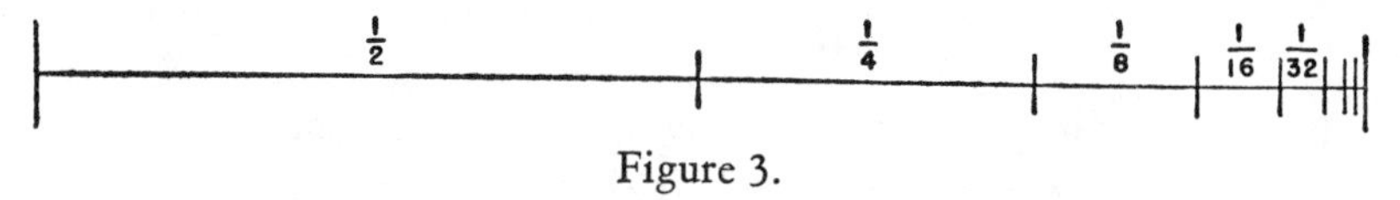

Figure 3.

half again. Keep repeating the bisections, as shown. Then, adding the pieces from left to right, you get

$$\tfrac{1}{2} + \tfrac{1}{4} + \tfrac{1}{8} + \tfrac{1}{16} + \tfrac{1}{32} + \tfrac{1}{64} + \tfrac{1}{128} + \cdots$$

Now this sum can never exceed 1, no matter how many terms you take, because there was only one foot of line there to begin with. In fact we say 'the sum to infinity *is* one.' You may object that the end of the line is never attained, that therefore the sum is actually always less than one. To answer this objection satisfactorily, a precise definition of limit would have to be given. It turns out that this matter of not reaching the limit, although it cannot be dismissed lightly, may not be important in a given problem. What is important is that the limit is *approached*, but not *exceeded*.

Consider a bouncing ball that moves through half a foot on its first bounce, a quarter of a foot on its second bounce, and so on forever. No real ball is so perfectly elastic that it would do this, even in a vacuum. But what if it did? Your answer must be only that the poor ball would bounce on and on until it disintegrated from fatigue, without ever quite traversing the distance of one foot. And you would be right—unless you took into account the time element. *How long* does each bounce take? If each bounce takes one second, the ball would never come to rest. But suppose the first bounce takes $\frac{1}{2}$ second, the second bounce takes $\frac{1}{4}$ second, the next $\frac{1}{8}$ second, and so on? Where is the ball at the end of one second?

This is the kind of analysis you must use if you are to worry your way out of Zeno's famous paradox of the tortoise and

Achilles, which has troubled mathematicians for centuries. Achilles, running to catch the tortoise, must first reach the point from which the tortoise started. But by that time, T has departed to a new spot. A then runs there, but T has just left. This goes on indefinitely, so that A can never catch T. Or can he?

• • •

How does our number system work? What do 4 and 7 have to do with 47? When we write 'xy' in algebra, we mean 'x times y.' When we write 47 in arithmetic, we certainly do not mean 4 times 7. We are so used to it that we sometimes forget just what we do mean:

$$47 = 4 \times 10 + 7.$$

Similarly,

$$3 \times 100 + 4 \times 10 + 7 + 2 \times \tfrac{1}{10} + 6 \times \tfrac{1}{100}$$

is condensed into 347.26, surely a most useful abbreviation.

When we try to decimalize the common fractions, we run into a problem. $\frac{1}{2} = .5$, and $\frac{1}{8} = .125$, arrived at by dividing the numerator by the denominator by actual long division, as you learned it from Miss Montgomery in the fifth grade. But what about $\frac{1}{3}$? When we try to perform the division we get

$$\tfrac{1}{3} = .33333 \cdots,$$

a non-terminating decimal (not to be confused with the non-terminating continued fraction we recently encountered, a very different critter). I dare to write the equals sign because you have already admitted (I hope) that a sum which is infinitely long to write can be finite in value. We will prove it again in a moment.

But although $.33333 \cdots$ does not terminate, at least it has the kindness to be *periodic*. We now undertake to show that this is characteristic of rational fractions. Every periodic decimal represents some rational number, and conversely, all rational numbers have periodic decimal expansions. A number like $2.7613613613 \cdots$ is called periodic, even though the periodicity does not begin until after the 7. Also numbers that 'come out even,' such as .5, are said to be periodic because they can be written $.50000 \cdots$ with the zeros repeating.

How much do you remember about geometric progressions?

$$.3333\cdots = \tfrac{3}{10} + \tfrac{3}{100} + \tfrac{3}{1000} + \tfrac{3}{10000} + \cdots$$

This is a geometric progression, or series. Its first term, a, is $\tfrac{3}{10}$. Each subsequent term is obtained by multiplying the previous one by $\tfrac{1}{10}$, called the ratio, r. Then an easily proved theorem of high school algebra, valid for all $|r| < 1$, tells us that the sum to infinity is

$$S = \frac{a}{1 - r} = \frac{\tfrac{3}{10}}{1 - \tfrac{1}{10}} = \frac{3}{10} \div \frac{9}{10} = \frac{1}{3}.$$

Any recurring decimal can be written in the form of a constant plus a geometric progression. For instance,

$$2.7613613\cdots = 2 + \tfrac{7}{10} + \tfrac{613}{10000} + \tfrac{613}{10000000} + \cdots$$

The progression part can be collected into a single fraction by the formula for S, and the result added to 2 and $\tfrac{7}{10}$ for the net rational number.

For the converse of the theorem, consider what happens when you decimalize $\tfrac{1}{7}$:

$$\begin{array}{r l}
 & .1428571\cdots \\
7 \,\big)\!\!\! & \overline{1.0000000\cdots} \\
 & \underline{7} \\
 & 30 \\
 & \underline{28} \\
 & 20 \\
 & \underline{14} \\
 & 60 \\
 & \underline{56} \\
 & 40 \\
 & \underline{35} \\
 & 50 \\
 & \underline{49} \\
 & 10 \\
 & \underline{7}
\end{array}$$

At first, each step of the division yields a different remainder: 7 into 10 yields a remainder of 3, into 30 a remainder of 2, and so on until at the sixth division we get a remainder of 1. Thus the seventh division is 7 into 10, which we started with. The whole process repeats, and from there on it *has* to become periodic. To summarize: when you divide by 7, there are only six *possible* different remainders. In general, by the same reasoning, dividing by n can give only $n - 1$ possible different remainders. Does this make it clear to you that the length of the longest possible period of any fraction is $n - 1$, where n is the denominator of the fraction?

We know that $\sqrt{2}$ is not equal to any common fraction p/q. Therefore its decimal expansion can never be periodic, however far the places are carried out. The decimal approximation for $\sqrt{2}$ to ten places is 1.4142135623, and of course it can be taken as far as you please by the ordinary arithmetic method of determining square root; but it will never show a period.

• • •

Look again at the first six numbers of the decimal expansion of $\frac{1}{7}$. Where have you seen this sequence of digits before? On the first line of this chapter. What is the decimal expansion of $\frac{2}{7}$, which is equal to $2 \times \frac{1}{7}$? Well, it starts at a different digit, but as soon as it gets rolling it looks just like $\frac{1}{7}$. Likewise $\frac{3}{7}$, $\frac{4}{7}$, $\frac{5}{7}$ and $\frac{6}{7}$. Even $\frac{7}{7}$ falls into line, because $.999999 \cdots = 1$. It is now clear why 142857 behaves the way it does. We can also answer the question, what other fractions besides $\frac{1}{7}$ produce cycles? Exactly those fractions $1/d$ whose decimal expansions have period of length $d - 1$, the maximum possible period.

Let us investigate the meaning of this answer.

• • •

A prime number is one which has no divisors except itself and 1. Thus 7 is a prime. We might guess that all primes (except 2 and 5, which are special because they divide 10 and hence 'come out even') have reciprocals * with decimal expansion of maximum

* The reciprocal of x means $1/x$.

period. This guess would be wrong. We have already met a prime reciprocal, $\frac{1}{3}$, which has period 1, whereas 2 would be its maximum possible period. 11 is the next prime after 7, but $\frac{1}{11}$ has period of length 2:

$$\frac{1}{11} = .090909 \cdots$$

$\frac{1}{13}$ has a period of length 6, not 12. The next prime is 17, and $\frac{1}{17}$ does have a period of length 16. Thus the number

.0588235294117647

is the next smallest number having the cyclic property.

There are only seven other numbers under 100 (besides 7 and 17) whose reciprocals have maximal period. These are 19, 23, 29, 47, 59, 61, and 97.

There is no easy way known at the present time of predicting the length of the period for any fraction $1/d$. It is known, however, that the number of digits in the period is the least positive exponent e such that $10^e - 1$ is divisible by d, provided d is not a multiple of 2 or 5.* Thus 9 is divisible by 3, so $\frac{1}{3}$ has period of length 1; 99 is divisible by 11, so $\frac{1}{11}$ has period of length 2. This means that 17 will divide evenly 9999999999999999, but no smaller number made up wholly of 9's. In order to find the length of the period of $1/d$, we need only divide d into a number consisting of 9's until it comes out even. But this is no harder or easier than dividing it into 1 until it repeats. The cure is neither better nor worse than the disease.

We have 'gone about as far as we can go' in our investigation of cyclic numbers. It has taken nearly a chapter to answer a seemingly innocent question. We were lucky: it might have taken a whole book or a lifetime. This is the invitation a mathematician accepts when he attacks any serious problem. You may not be satisfied with the answer. But again, we were lucky to get any answer at all. However tedious it may be to carry out a long division, at least *it can be done*. From the mathematical point of view, this constitutes a very definite solution.

* A consequence of a theorem of Pierre de Fermat (1601–65), probably the greatest mathematician of the seventeenth century.

We shall continue our short tour through the land of numbers by taking a further look at the primes.

• • •

Prime numbers have long fascinated mathematicians, professional and amateur alike. They have been studied since ancient times (Euclid showed that the number of primes is infinite). Yet comparatively little is known about them even to this day. There is a remarkable law dealing with their approximate distribution among the whole numbers. They get more scarce as you go out along the number sequence, but only gradually and very irregularly. There are

25 primes between 1 and 100
21 primes between 100 and 200
16 primes between 200 and 300
16 primes between 300 and 400
19 primes between 400 and 500
14 primes between 500 and 600
and so on.

The law of *approximate* distribution is given in the appendix. There is no known law saying *exactly* how the primes are distributed; perhaps none exists. We know so little about primes that we scarcely know what to look for. It is entirely possible that some deeply hidden property, when discovered, will unlock many secrets about primes. But many mathematicians through the centuries have spent thousands of hours searching for such a property, with only desultory results.

Here is an extraordinary fact about primes that is (for a change) easy to prove: even though the number of primes is infinite, and they are more or less 'scattered everywhere' among the numbers, it is nevertheless possible to find an interval *as large as you care to name* which is totally devoid of them. For instance, if you demand a sequence of consecutive whole numbers 1000 units long, *none* of which is a prime, I can produce it.

Consider the number *

$$1001! = 1 \times 2 \times 3 \times 4 \times \cdots \times 1000 \times 1001.$$

Now $1001! + 2$ is certainly divisible by 2, because $1001!$ is.
Likewise $1001! + 3$ is certainly divisible by 3
$1001! + 4$ is certainly divisible by 4

. . . .

$1001! + 1001$ is certainly divisible by 1001.

Thus I have exhibited a sequence of 1000 consecutive integers all of which are guaranteed composite (non-prime).

Of course, $1001! + 2$ and its immediate successors are all whoppers: they have 2571 digits apiece.

This theorem hints at the sparseness of the distribution of primes in the super-astronomical outer reaches of the number system.

• • •

If you want to make your name famous in mathematics, find a formula, or some kind of workable procedure, that will yield all the primes. There must be *some* law of formation of these special numbers; but if there is, it has eluded the best minds in the field to date. There is still basically no better way of finding whether a large number is prime than to try dividing it by all lesser primes, up as far as the square root of the number.**

Mathematicians would gladly settle for much less than a formula for *all* the primes: they would be happy to know of one that would be guaranteed to produce *some*. The meaning of this statement is best indicated by two examples. The formula

$$X = n^2 - n + 41$$

will yield a prime X for all integral n less than 41. But if $n = 41$, $X = 41^2$, composite. What is desired is a formula that will yield

* Read 'One thousand and one factorial.' The exclamation point does not express surprise, although there is a fair share of it in this theorem.

** Do you see why we need try only primes, and why only up to the square root?

a prime for *any* n used. Fermat thought he had such a formula:

$$X = 2^{2^n} + 1.$$

The first five 'Fermat numbers,' obtained by substituting $n = 0$, 1, 2, 3, and 4, are 3, 5, 17, 257, and 65537. They are all primes. But the next one,

$$2^{2^5} + 1 = 4294967297,$$

is composite,* and so are many subsequent ones. The moral is that one swallow, or even five, does not make a mathematical summer: it must be all or nothing. Fermat offered no proof of his conjecture (since, naturally, he was unable to find one). He only said he *thought* it was true. We shall see the importance of the distinction in just a moment.

• • •

A Diophantine equation (after Diophantus, of the third century of the Christian era) is one that demands a solution in whole numbers only. Thus the equation

$$x^2 + y^2 = z^2$$

has many whole-number (integral) solutions, the most familiar being $x = 3, y = 4, z = 5$. In fact there are infinitely many different x, y, z which will satisfy this Diophantine equation, and the formula that yields all of them is well known.

One naturally asks next, what about

$$x^3 + y^3 = z^3?$$

What integral x, y, z satisfy this equation? The answer is, none. This is the famous 'last theorem' of Fermat. The theorem is much broader: not only are there no such triples of cubes, but none of 4th's, or 5th's, or *any* higher power. The equation

$$x^n + y^n = z^n$$

* Its factors, discovered by Euler in 1732, are 641 and 6700417.

cannot be satisfied for any integral x, y, z with n an integer greater than 2.

Fermat did not publish a proof of this theorem, *but he said he had one*. This takes it out of the class of the conjecture described in the last section. All the other theorems which Fermat claimed he could prove have since been proved by other mathematicians—but not this one. It is known to be true for very large classes of n's; and no one has ever succeeded in finding a counter-example. But a complete proof is still lacking.* Although the generations of mathematicians who have spent heartbreaking hours in quest of it have no doubt roundly cursed Pierre de Fermat for his reticence, it was nevertheless a good thing for mathematics. The effort expended has been far from wasted: a by-product was the creation of a large part of the important Theory of Algebraic Numbers.

• • •

'For, were there means of doing so
They would have proved it long ago.'

De Morgan quoted these lines in the course of showing that the idea implied by them, however popular, is fallacious. The fact that 'they' haven't proved it yet does not necessarily mean that they never will. The 'means of doing so' may be hidden just around the next corner, waiting to be found by a gifted investigator with the necessary inspiration, plus mathematical background, plus downright hard work.

We mention two outstanding problems about primes which so far have stubbornly resisted the most determined attacks of the number people.

1. *Goldbach's conjecture:* every even number can be represented

* In June 1993 the mathematical world was electrified by the announcement that the famous theorem had at last been proved. Presentation of the outline of the solution required three lectures delivered by Andrew Wiles, a Princeton mathematician, who had been at work on the problem for seven years. Although his proof is so abstruse that it will take highly specialized mathematicians many months to check it, the experts agree that it looks very much as if the proof is valid. [1994 note.]

as the sum of two primes. Thus 8 = 5 + 3, 48 = 29 + 19, and so on. No even number has ever been found for which this is not possible. Yet no proof is known to date (1955), although progress has been made on the problem.

2. *There are infinitely many prime pairs.* A prime pair consists of two prime numbers whose difference is 2, like (11,13) and (29,31). They seem to be scattered throughout the number system. The statement that there are infinitely many of them is believed to be correct; but, as Courant and Robbins bluntly put it, 'not the slightest definite step has been taken toward a proof.'

3 / *What Are the Chances?*

WHAT IS PROBABILITY? Horace Levinson, in his excellent book on the subject, comes up with a three-line definition after having carefully laid five pages of groundwork. I shall risk a somewhat wordier definition without any groundwork.

If a certain trial (such as tossing a coin) can produce 'success' or 'failure' (such as head or tail); and if a very large number of such trials is made; and if the ratio of successes to the total number of trials is tabulated; then the probability that any *one* trial will be a success is the limit approached by that ratio as the number of trials tends toward infinity.

This definition takes a lot for granted. Among other things, it assumes the existence of a limit. But most readers will agree, if they can wade through the technical jargon, that this is what we mean when we say that the probability of tossing a head on any one toss is $\frac{1}{2}$. We mean that a very large number of tosses will turn up *about* half heads; and that the more tosses we make, the closer to $\frac{1}{2}$ will that ratio tend to become. Incidentally, this definition also takes care of loaded dice, flanged coins, and the like. It is, of course, nothing but the standard empirical (rather than the mathematical) definition.

In a large number of tosses, there will surely be *runs* of different lengths. An important principle of probability can be paraphrased by the statement that 'the coin doesn't remember.' That is, even if there has just been a run of 10 heads, the probability that the next toss will be a head is neither greater nor less than usual: unless the coin is phony, it is still exactly $\frac{1}{2}$. Some gamblers do not appear to recognize this principle. After a long run of red at Roulette, very few will bet on red. But if the wheel is fair, red is exactly as likely as black to win the next play.

• • •

If you toss a coin you may get a head or you may get a tail. Let us represent this 50–50 situation by the numbers 1, 1.

If you toss a coin twice, you may get two heads; or a head followed by a tail, or a tail followed by a head; or two tails. Grouping the middle two together, we have 1, 2, 1.

If you toss a coin three times, you may get three heads; or two heads and a tail in any of 3 ways: *H–H–T*, *H–T–H*, or *T–H–H*; or two tails and a head in any of 3 ways: *T–T–H*, *T–H–T*, or *H–T–T*; or three tails. These possibilities can be summarized by 1, 3, 3, 1.

It turns out that all this information can be assembled in a compact form in a triangular array.

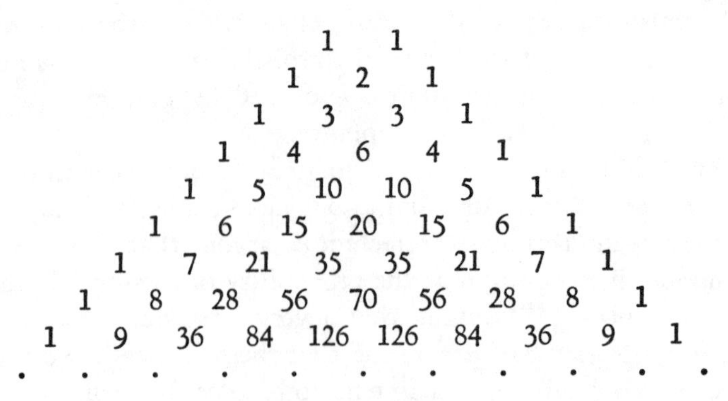

The Pascal Triangle

Each number in the triangle is obtained by adding the two nearest numbers from the row above. Thus in the seventh row, $21 = 6 + 15$, and so on.

Look now at the fifth row. It tells us that if 5 tosses are made, there is one way to get 5 heads, 5 ways to get 4 heads and a tail, and so on. Suppose we are trying to ascertain the probability of obtaining 4 heads and 1 tail in 5 tosses. There are 5 ways of doing this; and, by adding *all* the numbers in the fifth row, a total of 32 ways of tossing 5 coins. Thus the fraction $\frac{5}{32}$ describes the number of ways we can succeed divided by the total number of ways the experiment (of tossing 5 coins) can come out. We have arrived at the mathematical definition of probability.

The Pascal Triangle has many hidden properties. Among the most interesting is this: the first number after the 1 in each row divides all the other numbers in that row if and only if it is a prime.* Five and 7 divide all the other numbers in their respective rows, but 6, 8, and 9 do not. This is a general property of the whole triangle, however large you care to build it. So here we are suddenly back in Chapter 2. A triangular array of numbers, written down because it supplies information about probabilities, also answers the important question of whether or not a number is a prime. The connection, though not obvious, cannot be coincidental. We shall discover its nature in the next chapter.

We have found in the Pascal Triangle a criterion for primality; but—alas!—making the triangle big enough to test a large number is an even more arduous task than testing it by the old division method.

• • •

Your friend's family consists of 4 children, all girls, and a fifth baby is on the way. What are the chances that the new arrival will also be a girl? You should not bet more than even money on a boy. The odds are still 50–50, assuming that there are just as many girl-babies as boy-babies in a large population.

* This seems to be not well known. We give the proof in the Notes.

This is just another illustration of the principle stated in the opening section of this chapter.

You may voice the following objection: 'Consider all the families consisting of 5 children. Among these, surely there are not as many families of 5 girls as there are families of 4 girls and 1 boy?' That's right, there are not. In fact the Pascal Triangle will tell you just what the ratios are for all the combinations of 5 children. Here again is the fifth row:

1 5 10 10 5 1

For every one family of all girls you will find, on the average, about 5 families with 4 girls, 10 families with 3 girls, 10 with 2 girls, 5 with 1 girl, and one with no girls, all boys.

'Well, then,' you say, triumphantly, 'if there are five times as many 4–1 families as there are 5–0 families, and my friend's family already has 4 girls, why aren't the chances heavily in favor of the next child's being a boy?' The reason lies in the 'already.' A family that is already $\frac{4}{5}$ determined is no longer a random member of any 5-children survey; it is already heavily loaded with girls. More precisely, it is a question of *where* the boy fits into the family. Your friend's family cannot properly be compared with *all* the 4-girl, 1-boy families. It can be compared only with those 4-girl, 1-boy families in which the boy is the youngest child. And the number of families of *this* type is exactly the same as the number of 5-girl families. Hence the even-money proposition.

• • •

There is a famous problem in probability that is said to have trapped even skilled mathematicians into error from time to time. Yet it is easily stated, and appears innocent enough.

Three cards are identical in appearance except for their coloring, which is as follows: one card is red on both sides, one is white on both sides, and one is red on one side and white on the other. I shuffle them in a closed bag, and then reach in and draw one out and lay it on the table, without looking at or letting you see the side that is down. Suppose the side that is up is red. I then say, 'Obviously this is not the white-white card. Therefore it is either the red-white or the red-red. I'll bet you even money that it is the red-red.' If you take this bet and we repeat the game often enough, you will go home a lot poorer than when you came. The chances that it is the red-red are not even, but two to one in my favor.

This problem was stated and the solution above given in an article in the October 1950 *Scientific American* by Warren Weaver, the director of the natural sciences division of the Rockefeller Foundation. A spirited exchange of letters between Dr. Weaver and a professional gambler who challenged the correctness of the solution appeared in the December 1950 correspondence column of the same magazine. Who was right?

• • •

If I toss a coin 10 times, I may get exactly 5 heads and 5 tails, or I may not. But many people who are not very precise in their thinking reason as follows. 'Ten tosses are not enough. If you toss more times, you will stand a better chance of getting exactly half heads and half tails, "by the law of averages." And if you toss a very large number of times, say 1000, the chances are excellent that you will get exactly 500 heads and 500 tails.'

This is utter nonsense, and nothing of the kind happens. Old Pascal's Triangle again gives us the true story. Looking only at the even lines, because it is impossible to take exactly half of an odd number of tosses, we see that if we toss a coin just twice, the probability of getting half heads and half tails (one of each) is $\frac{1}{2}$. This probability is never so high again. From there on it goes down, not up! The probability of tossing exactly half heads in 4 tosses is $\frac{6}{16}$, or $\frac{3}{8}$ (six ways of succeeding out of sixteen ways of doing the tosses—the mathematical definition of probability). In 6 tosses, the probability of exactly half heads is $\frac{20}{64}$, or $\frac{5}{16}$. We display the first few probabilities in a table:

No. of tosses	*Probability of getting exactly half heads*
2	$\frac{1}{2} = .5000$
4	$\frac{6}{16} = .3750$
6	$\frac{20}{64} = .3125$
8	$\frac{70}{256} = .2734$
10	$\frac{252}{1024} = .2461$
12	$\frac{924}{4096} = .2256$

The interesting thing is that the probability keeps on decreasing toward zero. In mathematical language, it 'can be made arbitrarily small.' You can reduce the probability as far as you please by tossing 'a sufficiently large number of times.'

If you think about it, you will see that what it means, roughly, is this: as the number of tosses increases, there are more and more ways the number of heads can 'cluster around' 50 per cent heads without hitting 50 per cent exactly on the nose. What will happen in a very large number of groups, of 100 tosses in each group? Some will show 50 heads and 50 tails, to be sure. But almost as many will show 49 heads and 51 tails; and about the same number will show 51 heads and 49 tails. Not quite so many will show 48 heads; and another bunch, 52 heads; and so on. All these must 'get in their licks' to reduce the probability of exactly 50–50 under either the mathematical or the empirical definition (which, incidentally, reduce to the same thing for an unbiased coin).

• • •

Consider the following game of chance. We toss a coin, and if it comes down heads I pay you $2 and the game is over. If it comes tails, we toss again; and if this second toss is a head, I pay you $4, and the game is over. But if this second toss is a tail, we make a third toss, doubling the payoff again, and so on.

Now suppose we agree to play a large number of these games. Unless I charge an entry fee, I shall soon be broke. The question is, what is a fair price for you to pay to buy into each game? You will always get back at least $2, sometimes more, occasionally much more. It would seem that there should be some amount, the *average* of what I pay you over a large number of games, which would be a fair entry fee in the sense that, in the long run, we should about break even.

Mathematical expectation means the probability of winning an amount, multiplied by that amount. Suppose the game is simpler: we toss one coin once, and the game is over. If it comes down a head, I pay you $2. If it comes down a tail, I pocket your entry fee. In this case, everyone would agree that the entry fee should be exactly $1. Your mathematical expectation is also $1, because the probability is $\frac{1}{2}$ that you will win $2.

Now go to the next step. We play another game, slightly less simple. I pay you $2 if the first toss is a head; but if the first toss is a tail, we toss just once more. If now it comes heads, you get $4 and the game ends. If it comes tails, you get nothing and the game ends. This time your mathematical expectation is higher. Half the time you will get $2, a quarter of the time you will get $4, and the rest of the time you will get nothing. Hence your mathematical expectation is

$$\tfrac{1}{2} \times 2 + \tfrac{1}{4} \times 4 + \tfrac{1}{4} \times 0 = 1 + 1 = 2.$$

Your entry fee should therefore be $2. In half of the games you will get your $2 back. In the other half, the number of times you get $4 back will be exactly balanced by the number of times you get nothing back.

By the same argument, the entry fee for a game limited to three tosses is

$$\tfrac{1}{2} \times 2 + \tfrac{1}{4} \times 4 + \tfrac{1}{8} \times 8 + \tfrac{1}{8} \times 0 = 1 + 1 + 1 = 3.$$

But now what about an unlimited game? We are not going to stop until the run ends. Does this mean that the series continues indefinitely, and the entry fee should be an infinite amount of money?

Actually that is the correct answer. You will object that all runs must end *somewhere*, at a finite amount. But the question is, where? Where shall we cut off the series of 1's? If we could be sure there would never be a run longer than, say, 10, we could cut off the series there and the fee would be $10. But that is not the case. Sooner or later will come a run of 11 or 12 and upset the mathematical expectation, and also break the banker who tries to pay off. This is why the entry fee should, indeed, be infinitely large. You question the possibility of a run of very great length, such as 50? I say that if a sufficiently large number of games is played, there will at *some* time occur a run of length 50, paying 2^{50} dollars ($1,125,899,906,842,624). This run *might* occur today. That is why I cannot allow you to buy into the game for any finite sum.

If we agree to play the game for any finite entry fee, it can only be what Horace Levinson calls a risk unfavorable to both parties. For I must set the entry fee so high, in order to reduce my chances of being overtaken by disaster, that you would be foolish to buy in; and even so I am not fully protected. If I reduce the entry fee to an amount low enough to attract customers, I may ride the crest of the wave for a while; but now I am really courting ruination.

• • •

If ever you find yourself at a big party—of a dozen or more couples—which appears to be threatened by conversational doldrums, ask the assembled company whether they think it likely that any two of them have the same birthday. Since there are 365 different possible birthdays, most of the guests will say at once that it is extremely unlikely that any two birthdays coincide. Some of them will even be willing to place bets on it. I would take those bets. The chances are a little better than even that of 24 random people, at least two have the same birthday. And as the number of people involved increases beyond 24, the probability rises quite rapidly.

Try it. Don't be disappointed if it doesn't always work: it won't. But if you have an opportunity to do the experiment with several different groups, it will turn out more often than not that two birthdays coincide. If you are afraid they'll cheat, ask each one to write down his birthday on a slip of paper, and then collect the slips and read them off. Then it doesn't matter whether they write down the correct birthdays or not, as long as they don't know each other's: the dates on the slips will be random, which is all that is necessary.

A colleague of mine once offered this proposition to a large class of students, who received it with the usual skepticism. He then said, 'I'd be willing to bet that two of you in this very room have the same birthday.' The burst of laughter that greeted this remark puzzled him until he remembered the pair of identical twins sitting in the front row.

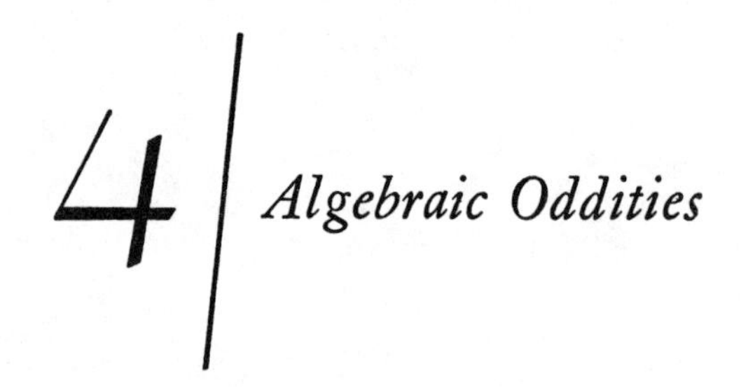

4 Algebraic Oddities

HERE is an old favorite algebraic fallacy. Where is the flaw?

Let	$a = b$
Multiply both sides by a:	$a^2 = ab$
Subtract b^2 from both sides:	$a^2 - b^2 = ab - b^2$
Factor:	$(a + b)(a - b) = b(a - b)$
Divide both sides by $(a - b)$:	$a + b = b$
But $a = b$; therefore	$2b = b$
Divide both sides by b:	$2 = 1$

Of course, the trouble is between the fourth and fifth lines. Since $a = b$, the quantity $(a - b)$ must equal zero. The fourth line is correct, the fifth line is not. We have broken what R. P. Agnew calls the fundamental commandment of mathematics: Thou Shalt Not Divide by Zero. Division by zero is against the rules of the game, and whenever you try it you will get something silly like $2 = 1$.

Here is another one, where the flaw (not division by zero this time) is more deeply hidden:

$$(x+1)^2 = x^2 + 2x + 1$$
$$(x+1)^2 - (2x+1) = x^2$$
$$(x+1)^2 - (2x+1) - x(2x+1)$$
$$= x^2 - x(2x+1)$$
$$(x+1)^2 - (x+1)(2x+1) + \tfrac{1}{4}(2x+1)^2$$
$$= x^2 - x(2x+1) + \tfrac{1}{4}(2x+1)^2$$
$$[(x+1) - \tfrac{1}{2}(2x+1)]^2 = [x - \tfrac{1}{2}(2x+1)]^2$$
$$(x+1) - \tfrac{1}{2}(2x+1) = x - \tfrac{1}{2}(2x+1)$$
$$x + 1 = x$$
$$1 = 0$$

¿Qué pasa?

• • •

To qualify for a certain auto race, each car must complete two laps around a one-mile track at an average rate of 60 miles per hour. One of the cars has engine trouble, and the driver finds that at the end of one lap he has averaged just 30 m.p.h. so far. What average must he maintain over the second lap in order to qualify?

You have read enough of this book by now to suspect that the answer is not the obvious one of 90 m.p.h. Why not?

• • •

A beekeeper who has sold one of his hives finds that he can best move it without disturbing the bees if he puts it in his open truck and drives slowly at a steady 6 m.p.h. to the buyer's house, which is at Centerville, 10 miles away down a straight road.

Just as he starts out, a flower vendor with a pushcart leaves Centerville and moves toward the bee truck at 4 m.p.h. An industrious bee leaves the hive at the start of the trip, flies to the pushcart in the proverbial bee-line, dips into a flower, and returns straight to the moving hive to deposit the goodies. The bee then makes another round trip, and repeats the process over the ever-decreasing distance until the truck and the pushcart meet. This is a very special kind of bee found only in mathematics problems: it takes no time out either to collect the pollen or to deposit it. The bee's velocity of flight is 8 m.p.h. How far has it flown when the truck and the pushcart come together?

Put that pencil away! This one can be solved easily and quickly in your head.

• • •

We have already encountered two kinds of infinite forms: continued fractions and geometric series. A third kind is represented by the expression

$$\sqrt{2+\sqrt{2+\sqrt{2+\sqrt{2+\cdots}}}}$$

The neatest way to evaluate such an expression is to observe that if it has any limit, say x, then x must be such that

$$x = \sqrt{2+x} \qquad \text{(Why?)}$$

Solving this equation by squaring both sides, we have

$$\begin{aligned} x^2 &= 2 + x \\ x^2 - x - 2 &= 0 \\ (x-2)(x+1) &= 0 \\ x &= 2, \text{ or } -1. \end{aligned}$$

But -1 is a so-called extraneous solution, not satisfying the original equation; so 2 is the limit of the iterated radical.

If we cut off this iterated expression at any finite stage, its value is clearly irrational; for it contains $\sqrt{2}$ at the end, and the square root of 2 plus an irrational number is still irrational. In Chapter 2 we found that an irrational number, $\sqrt{2}$, could be expressed as the limit of a sequence of rational numbers. Here we have a rational number, 2, expressed as the limit of a sequence of irrational numbers.

• • •

Why does minus two times minus five give plus ten? Attempts at justifying this rule include illustrations such as the following. It costs the state \$5 per day to feed, house, and clothe each prisoner at a penitentiary. Two convicts escape. Hence the prison counts -5 dollars, times -2 prisoners, and shows a profit of $+10$ dollars on its books for every day the fugitives are at large.

Such an illustration *proves* nothing. The true state of affairs is not in the least mysterious. If a certain law of multiplication of positive numbers (the Distributive Law) is to hold for negative numbers, too, then the rest follows logically. If you have never heard of the Distributive Law, you should be none the less willing to accept the following proof:

$$\begin{aligned}(-2)(-5) &= (-2)(-5) + (0)(5)\\ &= (-2)(-5) + (-2 + 2)(5)\\ &= (-2)(-5) + (-2)(5) + (2)(5)\\ &= (-2)(-5 + 5) + (2)(5)\\ &= (-2)(0) + (2)(5)\\ &= (2)(5)\\ &= 10\end{aligned}$$

Any numbers or letters can be used in place of 2's and 5's, to show that

$$(-a)(-b) = ab.$$

Many of the other mysteries of high-school algebra would never have been mysterious to you had they been properly explained at the outset. It is a pity that so many people learn algebra—or try to learn it—by memorizing a lot of things that the teacher says are so, without the slightest attempt at understanding *why*. You may remember, for example, that somebody once told you that anything raised to the zero power equals 1:

$$a^0 = 1.$$

Yet I doubt if one reader in 20, except for professional mathematicians, has the least conception of *why* this should be so. The reason is more obvious than that of the previous paragraph. If we wish our laws of exponents to remain in force, even if the result is zero, the definition $a^0 = 1$ is the only possible one:

$$\frac{a^5}{a^2} = \frac{a \cdot a \cdot a \cdot a \cdot a}{a \cdot a} = a^3,$$

by cancellation. Thus we see that when we divide powers of like bases, we arrive at the correct result by subtracting exponents:

$$\frac{a^5}{a^2} = a^{5-2} = a^3.$$

Apply this to a special case:

$$\frac{a^3}{a^3} = a^{3-3} = a^0.$$

But we already know that

$$\frac{a^3}{a^3} = 1,$$

by cancellation. Hence $a^0 = 1$.

Can you show by a similar line of attack why a^{-2} should mean $1/a^2$?

End of today's lesson. Class dismissed.

• • •

The Distributive Law used in the preceding section is the law of multiplication which allows us to write

$$2(x + y) = 2x + 2y.$$

The multiplication *distributes* the 2 over the x and the y. Observe that this is not the case with exponents:

$$(x + y)^2 \text{ does not equal } x^2 + y^2.$$

In fact, $(x + y)^2$ means take the expression $x + y$ and multiply it by itself:

$$\begin{array}{l} x + y \\ x + y \\ \hline x^2 + xy \\ \qquad xy + y^2 \\ \hline x^2 + 2xy + y^2 \end{array}$$

That is, $(x + y)^2 = x^2 + 2xy + y^2$

Likewise, $(x + y)^3 = x^3 + 3x^2y + 3xy^2 + y^3$

$$(x + y)^4 = x^4 + 4x^3y + 6x^2y^2 + 4xy^3 + y^4$$

$$(x + y)^5 = x^5 + 5x^4y + 10x^3y^2 + 10x^2y^3 + 5xy^4 + y^5$$

These numbers look familiar. They are the elements of the Pascal Triangle again. If we observe that

$$(x + y)^5 = (x + y)^4(x + y),$$

the actual multiplication gives us an insight on the law of formation of the triangle. Here is a skeleton of the multiplication, with the signs and letters omitted:

$$\begin{array}{cccccc} 1 & 4 & 6 & 4 & 1 & \\ 1 & 1 & & & & \\ \hline 1 & 4 & 6 & 4 & 1 & \\ & 1 & 4 & 6 & 4 & 1 \\ \hline 1 & 5 & 10 & 10 & 5 & 1 \end{array}$$

If we let x stand for a girl and y stand for a boy, and adopt for the moment the convention that x^4y means 4 girls, 1 boy, and so on, then the expansion of $(x + y)^5$ gives us in compact form all the information we needed for our discussion of the family problem in Chapter 3.

Why? A more detailed study of the subject reveals that certain probabilities are best expressed as quantities called *combinations*. By mathematical induction, it can be proved that the coefficients of the binomial expansion (the numbers of the Pascal Triangle) are precisely these same combinations. The material of these last two statements is easy to develop, but it would take too long to do it here. It is part of the contents of some senior high-school algebra courses.

How do the prime numbers get into the act? That, too, is shown by breaking down the coefficients of the binomial expansion into their combinatorial components and working from there.

The expansion of $(x + y)^n$ is thus the missing link connecting the various properties of the Pascal Triangle encountered so disparately in Chapter 3. We shall meet this versatile fellow yet again much later in the story.

• • •

There are a few old faithful problems that keep turning up in all the books of mathematical pastimes and recreations. One of these is the wine-and-water-mixture puzzle.

A glass is half-filled with wine and another glass, the same size, is half-filled with water. A spoonful of wine is taken from the first glass and poured into the second, and the mixture stirred thoroughly. Then a spoonful of this mixture is taken from the second glass and poured back into the first. Is the quantity of wine now in the first glass greater or less than the quantity of water now in the second glass?

You can start a rousing good argument with this one—but only because nearly everyone tries to do it the hard way. It is an elementary illustration of something that often happens in more serious mathematics: the right attack 'breaks' the problem in a jiffy. It can be done by algebra in a few minutes; but it can be done in your head in no time at all.

The wine in the first glass is now diluted to an unknown extent by water. Suppose for a moment that this mixture happens to be 97 per cent wine and 3 per cent water. The 3 per cent of water came out of the other glass; and the missing 3 per cent of wine went into the other glass to replace it, it matters not how or when. They both stand at the same level. Hence it follows that the *rest* of the liquid in the other glass is water—exactly 97 per cent of its contents. So the answer is, neither greater nor less—the same. Note that 97 per cent may be replaced by any arbitrary

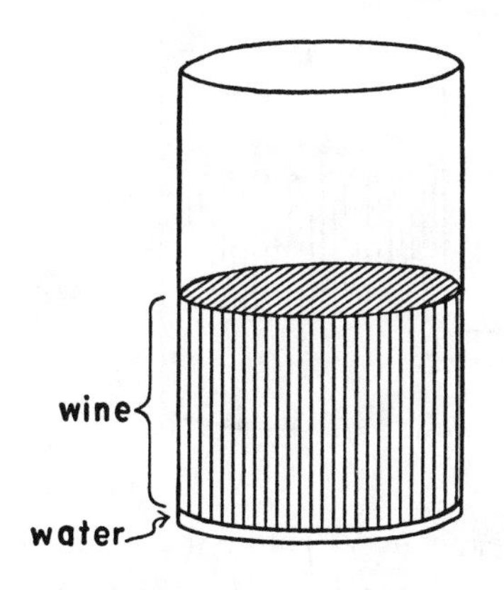

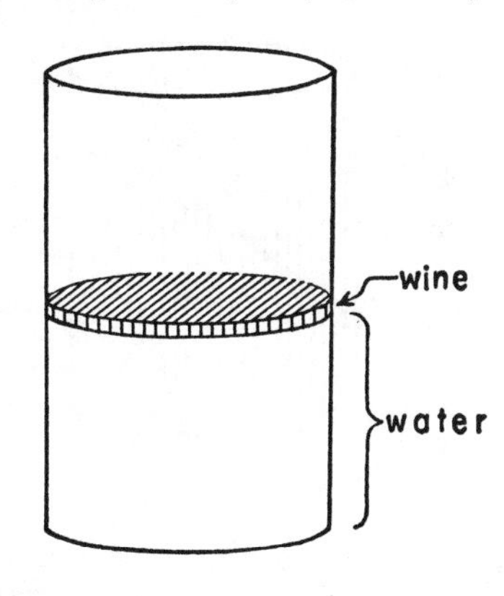

Figure 4.

per cent, and the result still holds. Some wine and some water merely swap places; and hence the respective amounts left behind are the same.

If this is not yet clear, look at Figure 4, which represents the situation after the spooning back and forth has taken place. You are to imagine, for the sake of the diagram, that the wine and water have somehow become separated. Note that the amount of water in the wine glass and the amount of wine in the water glass, although they are equal, do *not* represent one spoonful each. In fact, it doesn't matter how many times you repeat the operation with the spoon. You can make the problem sound more difficult by saying that it must be done twice; the answer to the original question will still be the same.

The problem can be generalized in an interesting way. Suppose that, instead of two glasses, there are two beakers with two con-

LIBRARY, UNIVERSITY COLLEGE CHESTER

necting pipes. The mixtures are forced continually through the pipes in opposite directions at the same rate by circulators, so that the same level is maintained in both beakers; and both mixtures are kept thoroughly stirred (Fig. 5). If the experiment starts with 100 cubic centimeters of pure wine in one beaker and 100 cubic centimeters of pure water in the other, and the circulators pump at the rate of one cubic centimeter per second, how soon will the wine and water be completely intermingled, so that a

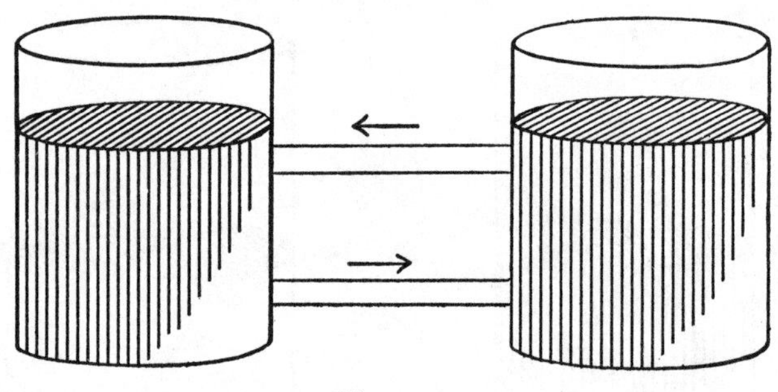

Figure 5.

sample from either beaker will test exactly 50 per cent of each?

The answer is never. The problem is easily solved by the methods of ordinary differential equations, a topic usually studied next after elementary calculus. The number of cubic centimeters of wine left in the wine beaker after t seconds is found to be

$$w = 50(1 - e^{-t/50}).$$

(Refer to the algebra lesson of a few pages back for the meaning of the negative exponent.) With the aid of this equation, we can tabulate values of w for any time t. Thus

after 50 seconds, $w = 68.4$ cubic centimeters
" 100 " $w = 56.8$ " "
" 150 " $w = 52.5$ " "
" 200 " $w = 50.9$ " "

The graph of w against t looks like Figure 6. After 600 seconds (10 minutes) w is approximately 50.0003. Of course this cannot be shown on the graph; it is far too close to the 50-line to be distinguishable from it. This happens after only 10 minutes. If the circulators were kept going for an hour, no known test would be sufficiently delicate to detect any difference in the percentages of wine in the two beakers, even if it were practically possible to

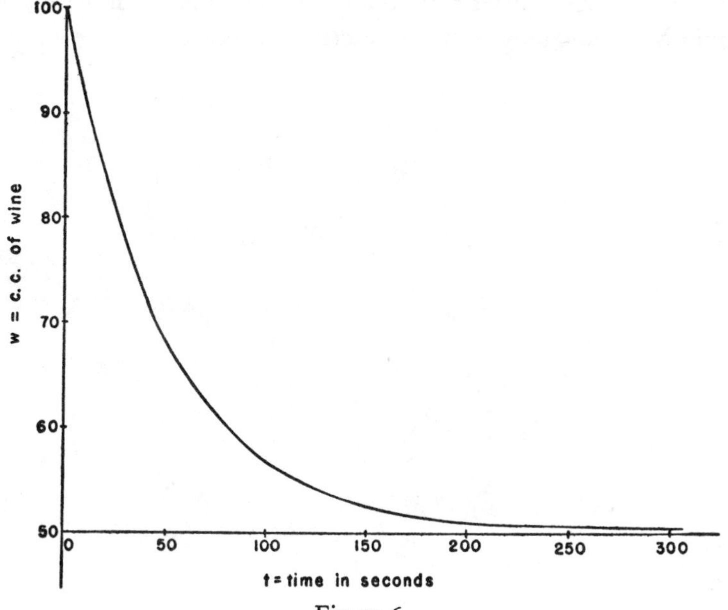

Figure 6.

maintain so precisely the theoretical mixing conditions. Eventually it would become a question of counting molecules, a discontinuous process. And so even the equation would finally, in the last analysis, have to be abandoned.

It would be safe to say, however, that at the end of some relatively short period, such as 10 minutes, the wine and water could be considered completely mixed 'for all practical purposes.'

• • •

A professor, asked what he meant by the last phrase in quotation marks, explained:

'Suppose all the young men in this class were to line up on one side of the room, and all the young ladies on the other. At a given signal, the two lines move toward each other, halving the distance between them. At a second signal, they move forward again, halving the remaining distance; and so on at each succeeding signal. Theoretically, the boys would never reach the girls; but actually, after a relatively small number of moves, they would be close enough for all practical purposes.'

5 / *Geometry, Plane and 3-D*

PROBABLY THE MOST FAMOUS THEOREM of plane geometry is the Pythagorean Theorem: the square on the hypothenuse of a right triangle is equal to the sum of the squares on the other two sides. The Greeks spoke of the square *on* rather than *of* the hypothenuse because they thought of the actual geometric square (Fig. 7) instead of the square of a number representing the length of a line. We have already used the theorem when we stated that the diagonal of the unit square has length $\sqrt{2}$ (Fig. 1).

The smallest integers that fit the Pythagorean relation are 3, 4, and 5:

$$3^2 + 4^2 = 5^2.$$

Is there any comparable geometric significance to the following less publicized relationship?

$$3^3 + 4^3 + 5^3 = 6^3.$$

If there is, I have never heard of it.

• • •

Of course *any* right triangle—not only those whose sides are whole numbers—obeys the Pythagorean Theorem. It has a vast

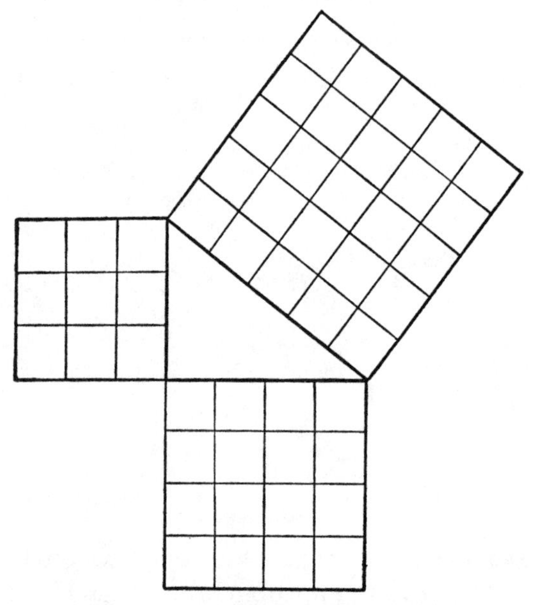

Figure 7.

number of different proofs. Every so often a 'new' one is published; but pretty soon along comes a letter to the editor, pointing out that the same proof can be found in some forgotten European textbook of the last century, where it is credited to Archimedes. W. F. White has aptly said, 'One who invents anything in elementary mathematics is likely to find that the ancients have stolen his ideas.' B. F. Yancey and J. Calderhead collected and classified 100 different proofs, some of which had many hundreds of variations depending on positions of triangles and choices of ratios.

One of the greatest mathematical works of all time is Euclid's *Elements*, written about 300 B.C. The vitality of the book is reflected by the fact that until recently it constituted (in translation from the Greek) the going text in plane and solid geometry. Schoolboys of only two or three generations ago spoke of studying Euclid when they meant geometry. The influence of the *Elements* is so strong that even today the best geometry texts use Euclid's proofs. This has one unfortunate result: occasionally his methods were tedious and difficult, and these difficulties have been perpetuated by the powerful Euclidean tradition. A case in point is the proof he used for the Pythagorean Theorem. In the light of the enormous wealth of proofs available today, it seems a pity to inflict schoolboys with one of the most awkward. Whether beautiful ones exist is for you to decide after looking at two samples. The diagrams speak for themselves (Figs. 8 and 9).

• • •

There are infinitely many solutions to the Pythagorean problem: find two (integral) squares whose areas added together equal the area of a third square. Besides $3^2 + 4^2 = 5^2$, we have $5^2 + 12^2 = 13^2$; $8^2 + 15^2 = 17^2$; and so forth.

What about cubes? Geometrically the problem would be to find two cubes whose volumes added together would equal the volume of a third cube—and all three must be cubes whose edges are integers. Algebraically, this is equivalent to solving in integers the equation

$$x^3 + y^3 = z^3.$$

We have seen that this is impossible, by Fermat's last theorem (Chapter 2). Furthermore, 3 is one of the values of n for which the theorem *has* been proved.

• • •

René Descartes (1596–1650) was the first to exploit the idea of solving geometric problems algebraically. He developed the theory of what in high school you called *graphs*. Let equally spaced intervals along a line called the X-axis be numbered 0, 1, 2, 3, $\cdots$ and the same along a line at right angles to it, the Y-axis.

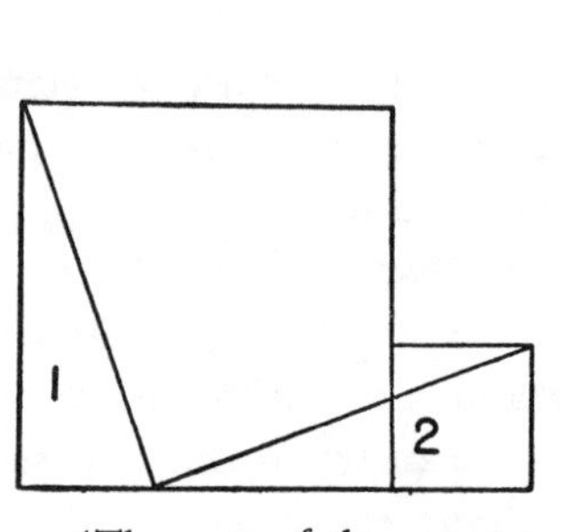

'The sum of the squares on the two sides . . .

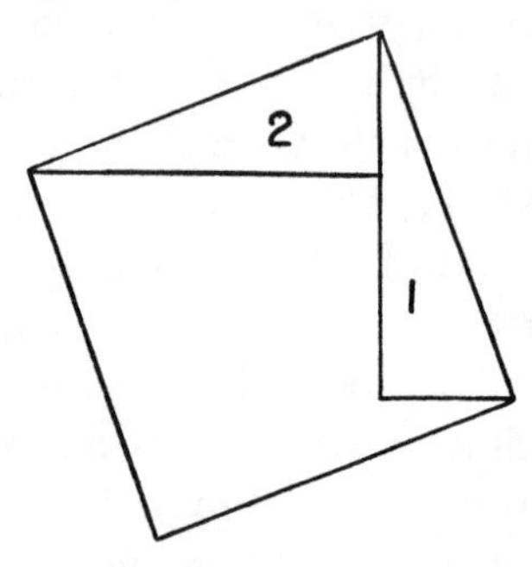

. . . equals the square on the hypothenuse.'

Figure 8.

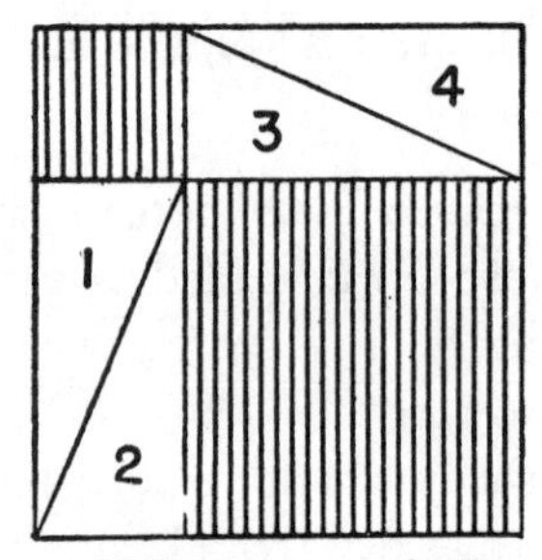

'The sum of the squares on the two sides . . .

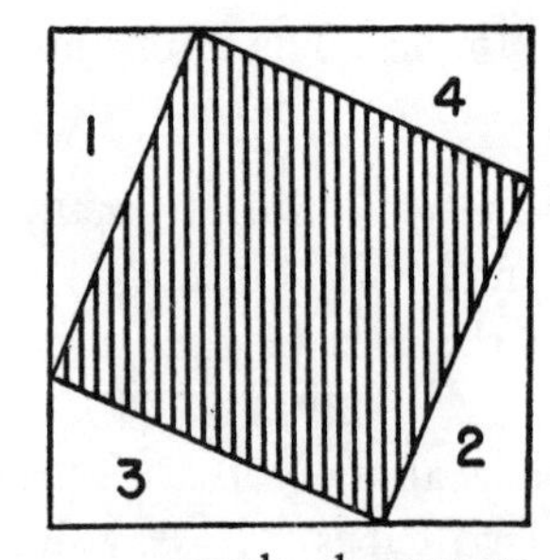

. . . equals the square on the hypothenuse.'

Figure 9.

Also negative integer spaces, as shown in Figure 10. Thus every point where two lines of the graph paper intersect is represented by two *co-ordinates*. The pair (3,2) represents the point P, 3 units to the right of the Y-axis and 2 units above the X-axis.

Now think of every point in the plane having such a representation—not just the whole-number points at the intersections of the lines. Then all the fractional points like $(\frac{3}{2}, -\frac{27}{53})$ would be there. Between two fractions, however close together they are, you can always insert another one (their average). Therefore all the fractional points constitute a very tight array: to put them all in would somehow 'fill' the whole plane. This is what the mathematician calls an *everywhere dense set*.

Yet, even though everywhere dense, the set of points both of whose co-ordinates are rational does not exhaust *all* the points

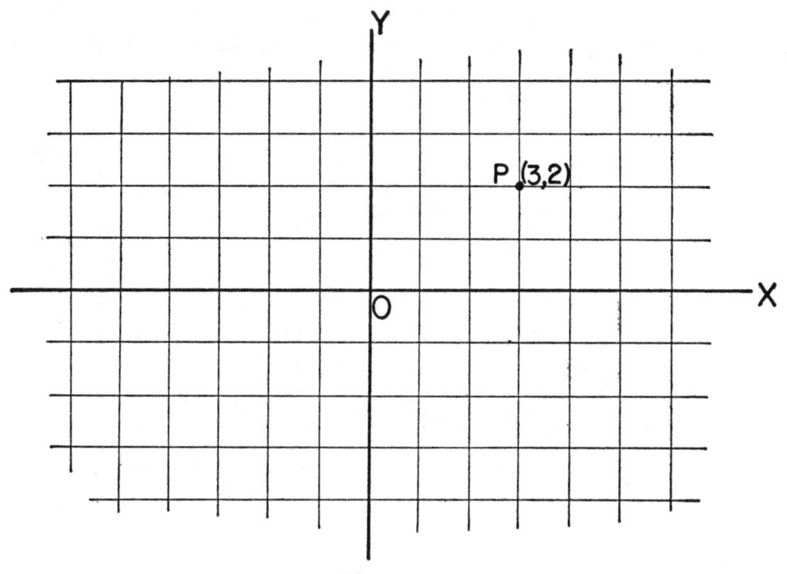

Figure 10.

of the plane; for we know that there are other points, such as $(\sqrt{2}, \sqrt{3})$, with one or both co-ordinates irrational. The irrationals form another everywhere dense set, interlarded among the rationals.

Fermat's last theorem, with $n = 3$, has a remarkable consequence which we can now describe. Since

$$x^3 + y^3 = z^3$$

has no solution in *integers*, we can surely say that

$$x^3 + y^3 = 1$$

has no solution in *rational* x and y other than (1,0) and (0,1).

For if it did, we could clear the fractions and get integers satisfying the Fermat equation. The graph of the equation

$$x^3 + y^3 = 1$$

means the curve all of whose points satisfy the equation. It looks like Figure 11. No points both of whose co-ordinates are rational, except (0,1) and (1,0), can satisfy the equation. Therefore, except for these two points, the curve of Figure 11 threads its way

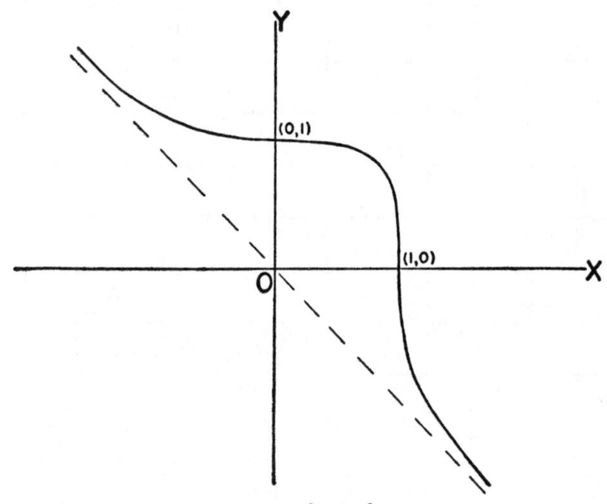

Figure 11. $x^3 + y^3 = 1$.

through the everywhere dense field of points both of whose co-ordinates are rational *without touching a single one of them.*

• • •

Imagine two solid cubes, made of steel if you wish, one just a trifle bigger than the other. It would be possible, with careful machining, to cut a hole or tunnel through the bigger cube large enough so that the smaller cube could be passed right through and out the other side. The big cube would be a mere shell of its former self, but you will grant that it could be done.

Now take a deep breath: it is also possible to cut a hole in the *smaller* cube big enough to pass the *big* one through! No tricks, either; it must be an honest hole, honestly surrounded by material of the smaller cube, and neither cube is to be compressed or stretched or deformed in any way.

Hold the page level, straight in front of you, and look at Figure 12. You see a regular hexagon, partitioned into six triangles. Now turn the page slightly until the word 'TOP' is horizontal, and look again. Do you see a cube?

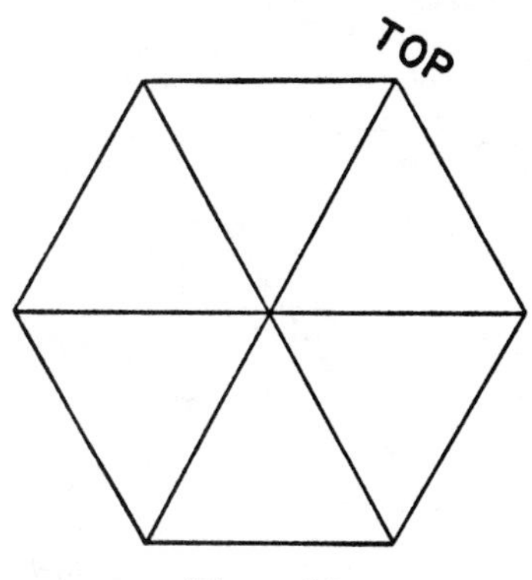

Figure 12.

Figure 13 is the same hexagon with certain lines left out. This hexagon is the actual outline of a cross-section of a cube through its center perpendicular to a main diagonal. If the cube measures one inch on an edge, there is no line in Figure 13 exactly one inch long; all the edges are foreshortened. But the straight-line distance from A to B, both on the drawing and in the unit cube itself, is $\sqrt{2}$ inches,

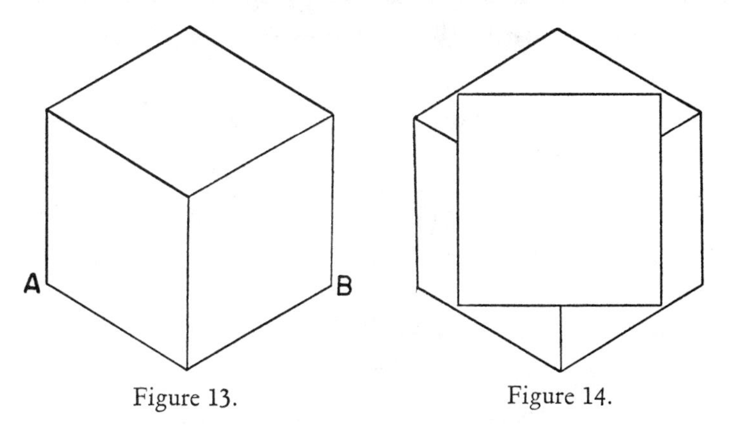

Figure 13. Figure 14.

since this distance is not foreshortened. Figure 14 shows a cross-section of a one-inch-square hole through the same one-inch cube. There is enough material left so that the hole could be enlarged a little more—not much, but just enough so that a cube slightly larger than one inch on an edge could slide through.

• • •

A *regular tetrahedron* is a triangular pyramid, its four faces (including the base as one face) all being equilateral triangles. A cross-section parallel to the base is evidently a smaller equilateral triangle. But by slicing judiciously, it is possible to get a *square* cross-section.

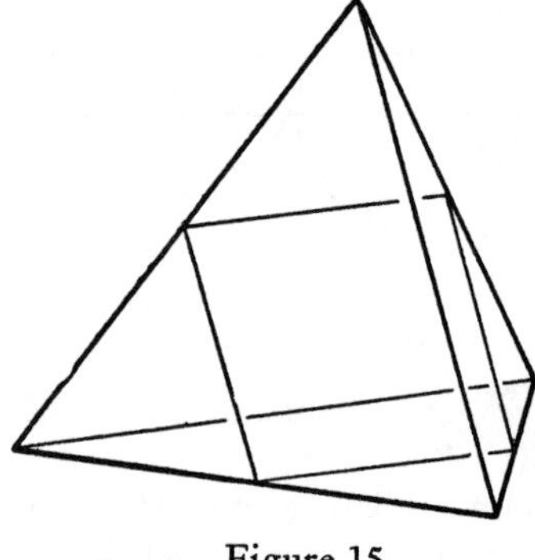

Figure 15.

Figure 15 shows the square, and describes better than words how to obtain it. It is a section parallel to two non-adjacent edges, midway between them. An actual plastic model of a tetrahedron cut at this plane has been available in toy stores under the trade name 'Magic Pyramid.' The two congruent solids (Fig. 16) do not in the least suggest their original orientation. Most people, confronted with the pieces and asked to reassemble them

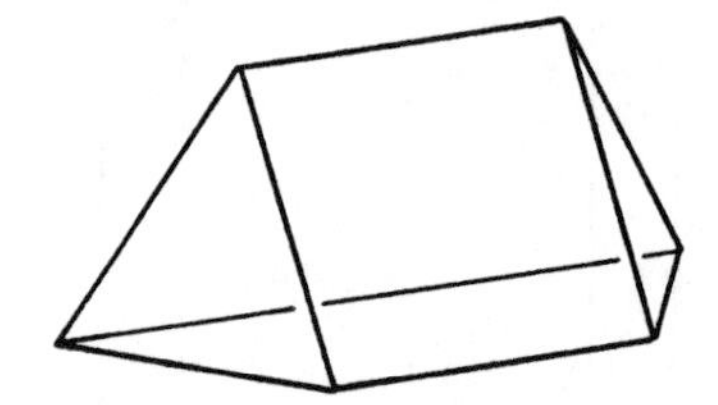

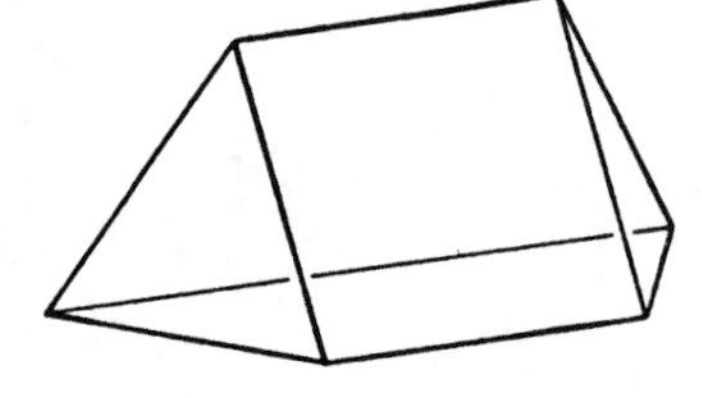

Figure 16.

to form a triangular pyramid, are unable to do so without several minutes' concentrated effort. It is quite surprising that such an extremely simple combination—two *congruent* halves of a *regular* solid—can afford any puzzle at all.

• • •

A *regular polyhedron* is a solid with congruent regular polygons for all its faces, and with its face angles (dihedral angles) all equal. We have just encountered two regular polyhedra, the tetrahedron and the cube, having respectively 4 and 6 faces. It might seem that there could be many more, perhaps an infinite number; but there are actually only three more, the ones with 8, 12, and 20 faces. These five regular solids were known to Plato, and are sometimes referred to as the Platonic solids.

How can we say with such assurance that there are only five regular solids? The proof is easy. It hinges on the fact that if we try to form any more, we run into difficulty at the corners.

The faces must be *regular* polygons. The regular hexagon has a vertex angle of 120°; all regular polygons of more than six sides have even greater vertex angles. It is impossible to fit three hexagons together (to begin a polyhedron) except in a plane; and it is impossible to fit three higher polygons together at all, for three vertex angles add up to more than 360°. This restricts us to the use of triangles, squares, and pentagons for the faces. The rest of the proof follows immediately. We can form a vertex with 3 squares or with 3 pentagons, which accounts for two solids, the cube and the dodecahedron. We can form vertices with 3, 4, or 5 triangles, since the vertex angle of an equilateral triangle is only 60°. These take care of the tetrahedron, octahedron, and icosahedron.

The five regular polyhedra are shown in Figure 17.

• • •

Here is a little puzzle that illustrates an important mathematical principle. If you can't solve it after studying Figure 18, you may turn to the Notes for the solution. The figure shows a cube made up of two solid blocks of wood securely dovetailed together. The appearance of the two vertical faces that are not shown is exactly the same as that of the two visible ones. The two pieces were not necessarily carved from the same block of wood; the sketch indicates that the grain runs in different direc-

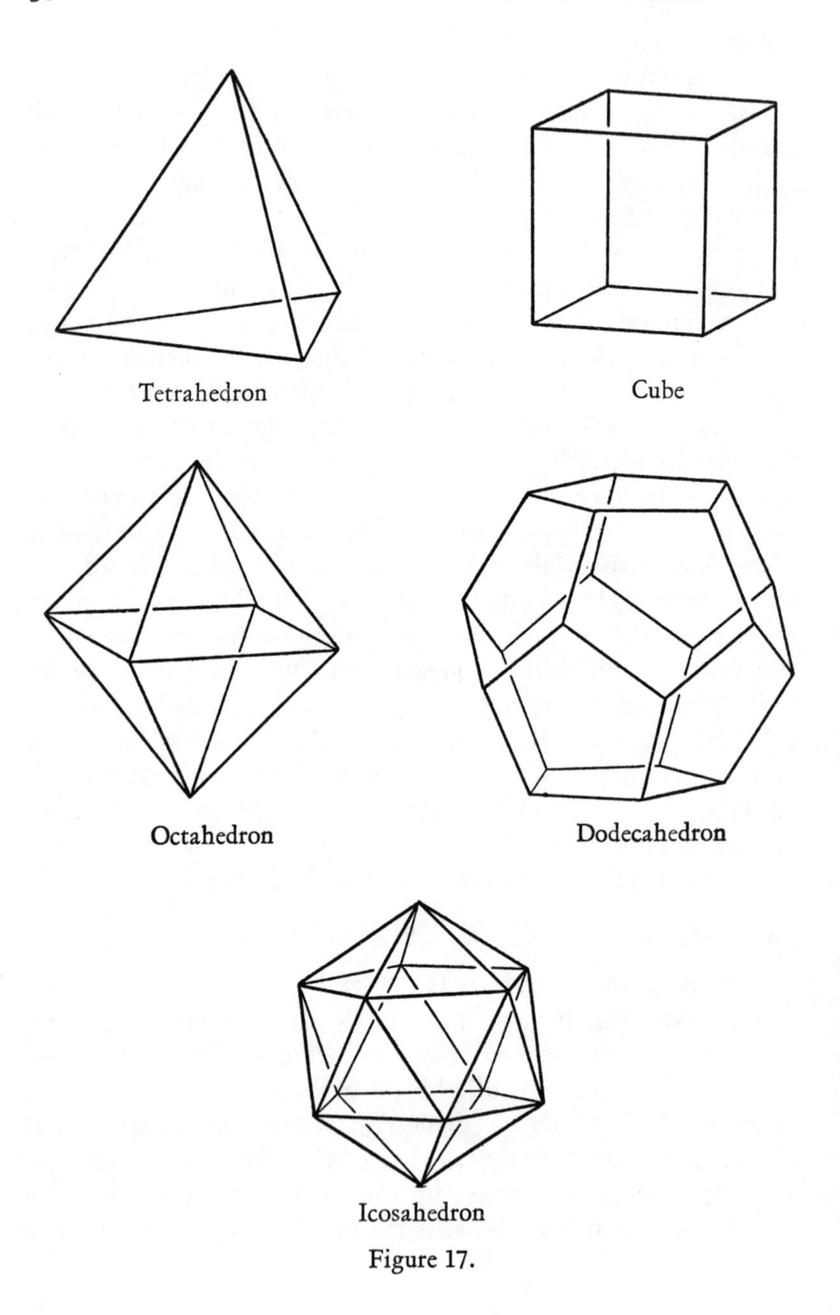
Tetrahedron
Cube
Octahedron
Dodecahedron
Icosahedron

Figure 17.

tions. The question is, how were the two pieces put together? No tricks are allowed: nothing is glued, nothing is steamed or compressed or forced into place, and there are no hollow spaces inside the cube.

Impossible, isn't it?

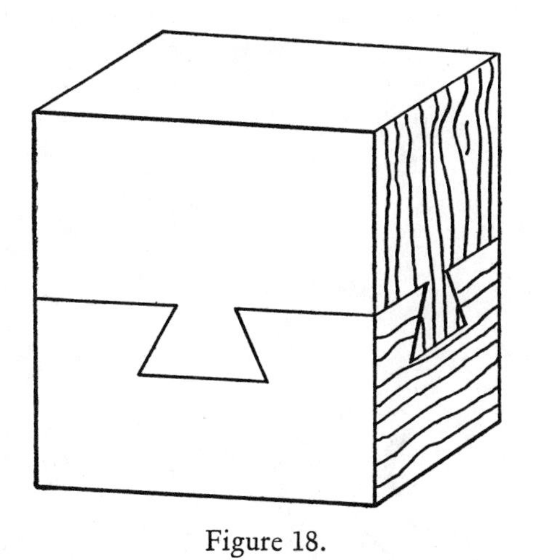

Figure 18.

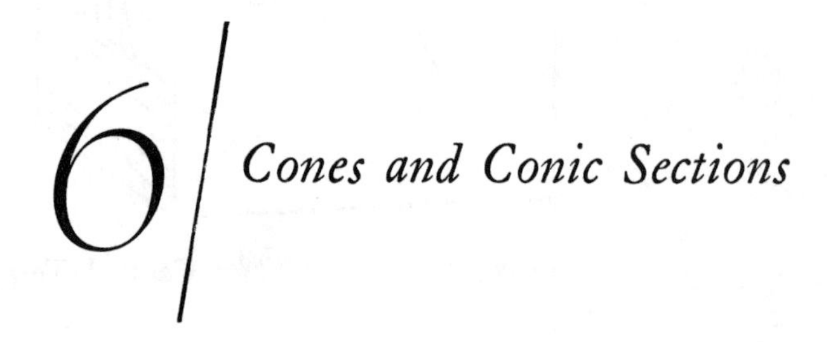

6 / *Cones and Conic Sections*

An ellipse is an oval—but not just any oval. It is the path of a point which moves (in a plane) so that the sum of its distances from two fixed points remains constant. An ellipse is easily drawn with the aid of two tacks and a piece of string, as shown in Figure 19. One end of the string is secured to each tack. Since the whole string does not vary in length, the path of the pencil conforms with the definition of the ellipse.

The location of each tack is called a *focus*, because of one of the properties possessed by an ellipse: light rays emanating from a source at one focus and reflected by the ellipse would all return to the other focus. This we prove by recalling a physical property of light: it travels always by the shortest available path. Suppose the tangent line m at any point P of an ellipse is a plane mirror (Fig. 20). Then there is some point of m at which light from F_1 is reflected to F_2 (namely, the point where a person at F_2 would

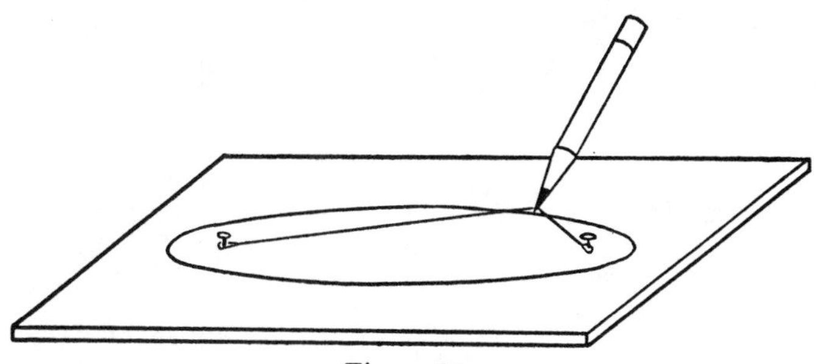

Figure 19.

'see' F_1 through the mirror). This point must be P itself; for the path F_1QF_2, via any other point Q, is *longer* than F_1PF_2 (the string would not reach to Q).

This reflecting property of the ellipse explains the sound effects obtained in certain so-called whispering galleries. The next time you are in New York City's Grand Central Terminal with a friend, you can make this experiment. As you start up the ramp from the lower level, just outside the Oyster Bar you will notice

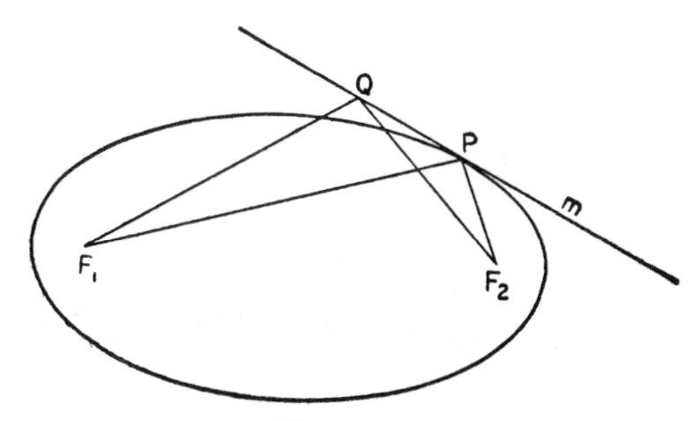

Figure 20.

an arched ceiling. If you and your friend stand close to and facing the wall, at opposite corners of the intersection of the two ramps, as at A and B in Figure 21, you can talk in very subdued tones and hear each other perfectly, while people passing between you

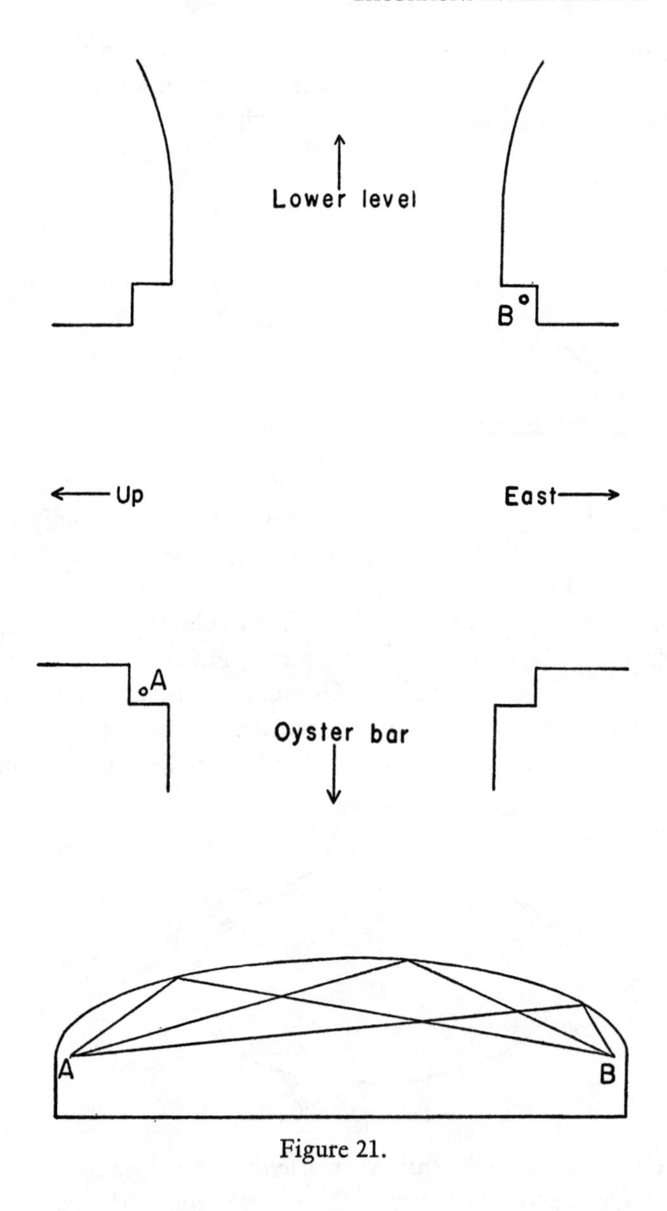

Figure 21.

will be unable to hear your conversation. You are standing at the two foci of an elliptical part of the ceiling.

• • •

The parabola, although frequently defined otherwise, can be considered as the limiting case of an ellipse whose foci have been moved far apart. If we fix one focus and one vertex * of the ellipse and move the other focus indefinitely far away, the ellipse tends toward a parabola (Fig. 22).

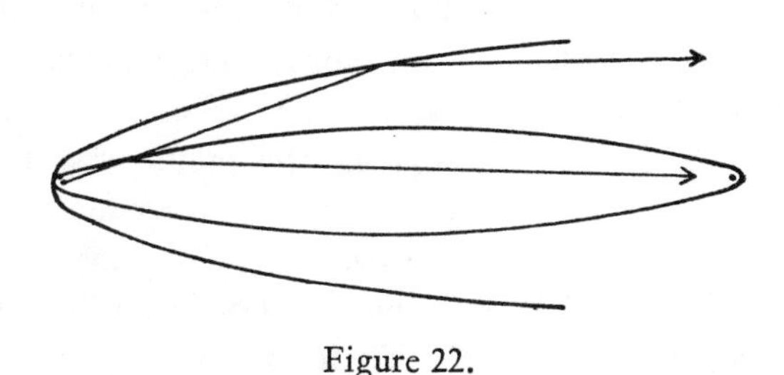

Figure 22.

Since light from the focus of a parabola is all reflected toward the 'other focus' infinitely far away, the light rays form a parallel beam. This is the principle of all parabolic reflectors. The form of the reflecting mirror in your automobile headlights or your pocket flashlight is a paraboloid of revolution (Fig. 23) with the bulb filament located at the focus. Reflecting telescopes make use of the same property, with the light traveling in the opposite direction. The 200-inch mirror of the Hale telescope at Mt. Palomar is a carefully engineered and painstakingly ground paraboloid that picks up light from the stars and reflects it to a focus 55 feet away, where it forms an image on a photographic plate. (This image can also be viewed through magnifying eyepieces, but it almost never is. A modern large telescope is really a giant camera for photographing the

Figure 23.

* The vertices of an ellipse are the points where the major axis, or longest 'diameter,' cuts the ellipse. *Vertex* here does not imply anything angular.

sky.) The same principle is used by parabolic radar screens to pick up electronic impulses.

• • •

The usual definition of the parabola is the path of a point which moves (in a plane) so that its distance from a fixed point equals its distance from a fixed line (Fig. 24). The fixed point is the focus. With this definition, it is an easy exercise in the calculus to deduce the reflection property. Is there a neat *anschauliche* proof, like the one in the previous section which does not need the calculus, that can be based on this more usual definition? I do not know of any.

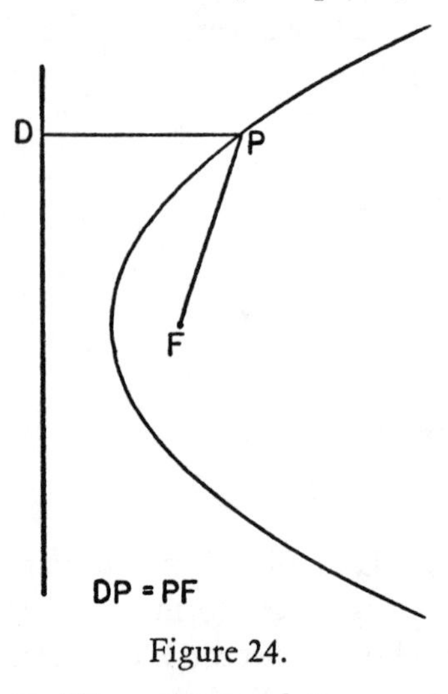

Figure 24.

There are mechanical devices for drawing parabolas, one of which is illustrated in Figure 25. One end of the string is fastened to a nail in the blackboard, or securely held, at *F*. The other end is attached to the end of the yardstick. The string is held tightly against the yardstick by the chalk, while the yardstick, kept vertical, is moved from side to side with its bottom end always at the same height (say, resting on the chalk-rack at the bottom of the blackboard). It will be seen that the yardstick 'measures out' additional bits of string, so that the distance from the chalk to the chalk-rack always equals its distance to *F*.

If you throw a stone, its path through the air is very nearly parabolic. The mathematics and physics textbooks state that, if the resistance of the air could be neglected, a projectile fired from a gun would follow a parabolic path. That this statement is inaccurate is not generally realized. Even in a perfect vacuum the trajectory of a bullet would not be parabolic but elliptical. The textbook error stems from the fact that gravity is assumed to act always in the same direction (straight down, of course). Actually gravity acts always toward the center of the earth, and thus its direction changes from point to point on the surface. My 'straight down' is not parallel to yours. For small trajectories this fact escapes unnoticed, because the difference between the elliptic and parabolic arcs is exceedingly small. But at White Sands, New

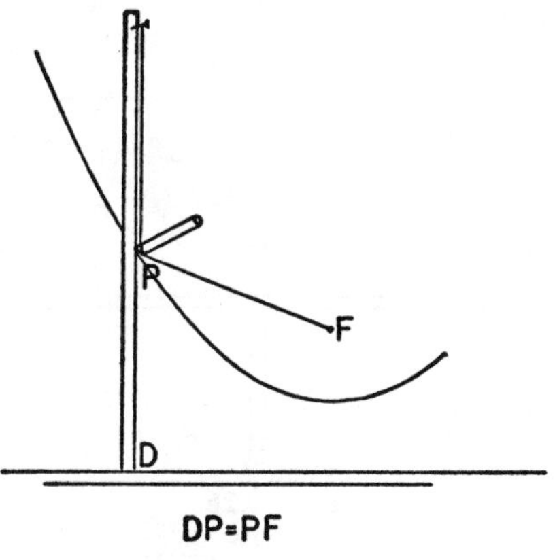

DP=PF

Figure 25.

Mexico, where rockets are shot for distances of 100 miles or more, the difference has a very substantial effect. The trajectory must be figured as a Keplerian orbit with the center of the earth as a focus. One assumes that the statement about parabolas will be eliminated, or at least qualified, in the textbooks of the future.

• • •

The hyperbola is defined as the path of a point which moves (in a plane) so that the *difference* between its distances from two fixed points remains constant. (Compare with the definition of an ellipse.)

A possible device for drawing a hyperbola is shown in Figure 26. The two strings are fastened independently to the pencil. They pass through holes in the drawing board at F_1 and F_2 and thence through an eyelet to the hand, which grasps them both together. As the hand moves up or down, it lets equal amounts of string through each hole and hence maintains a constant difference between the distances from the holes to the pencil. The strings must be interchanged to draw the other branch.

Hyperbolas find practical application in LORAN (long-range air navigation), a device that measures time differences between the receipt of two signals sent out simultaneously from two stations at far distant parts of the earth. The air navigator can thus locate himself somewhere on a hyperbolic curve. If he can receive another set of signals to provide an intersecting hyperbola, he has a 'fix.'

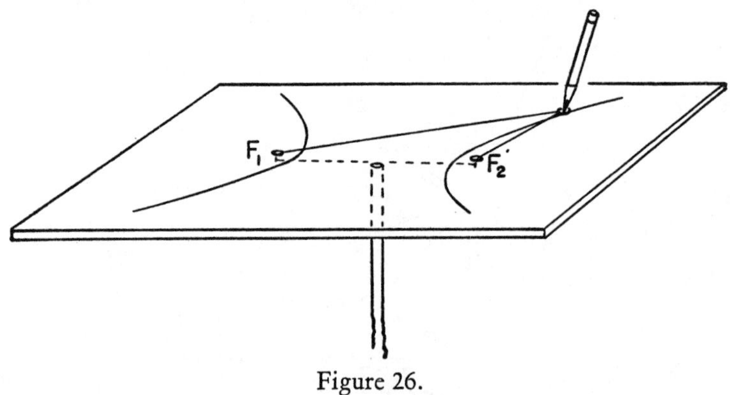

Figure 26.

• • •

The mathematical applications of the ellipse, the parabola, and the hyperbola are exceedingly numerous. Taken together with the circle (which is really nothing but an ellipse whose two foci have been brought together) they form a class of curves called *conic sections*. The conic sections all have equations in the Cartesian plane which are algebraic and of the second degree.* Why the name conic sections? Because each of these curves can be obtained by sectioning (slicing) a right circular cone with a plane. Figure 27 shows four of these sections. Any section parallel to an ele-

* The converse is not true, because some second-degree equations have imaginary curves. But except for these, all second-degree equations are conic sections or degenerate cases thereof.

ment of the cone is a parabola; any section cutting both parts (nappes) of the cone is a hyperbola.

It is all very well to say that these curves are the same ones we defined earlier as paths of points moving in certain restricted

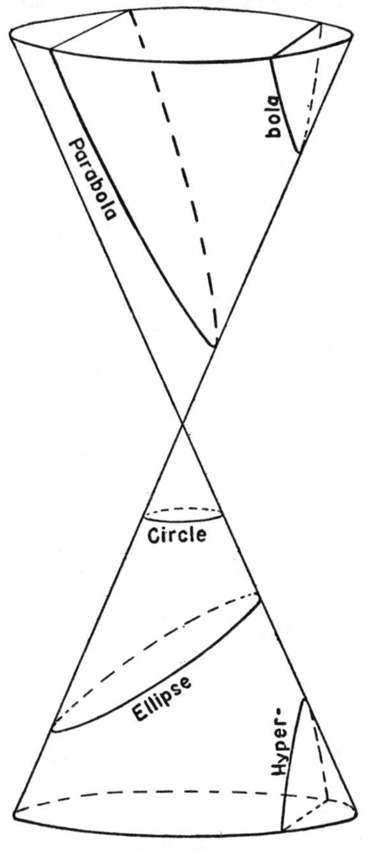

Figure 27. The conic sections.

ways. But how can we possibly defend such a claim? For example, it is not at all obvious that cutting all the way through one nappe must yield an ellipse. Maybe the curve is really egg-shaped, bigger at the lower end where the cone is wider?

The proof that it is not egg-shaped but actually an ellipse is so easy and so spectacular that we give it here, although in these days of abbreviated analytic-geometry courses, few college students ever see it. It was invented by G. P. Dandelin, a Belgian mathematician of the nineteenth century. The proofs for the other two kinds of sections go very similarly.

In Figure 28, we are trying to find out something about the

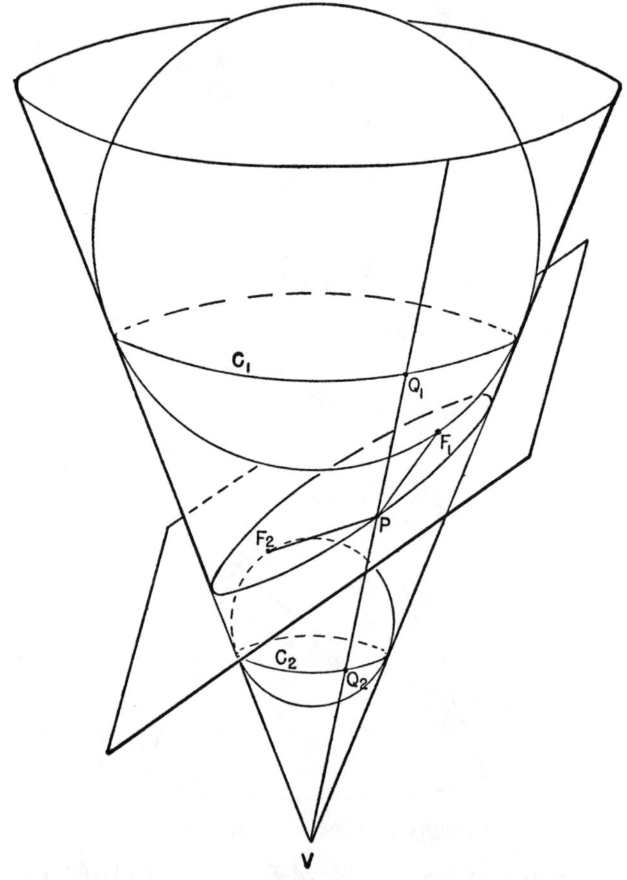

Figure 28.

oval which the plane cuts out of the cone. There is one sphere, and only one, which would fit into the cone, tangent to it along some circle C_1, and at the same time tangent to the upper side of the plane at some point F_1. There is just one other sphere which can be fitted into the cone in this way, and it is the one that is tangent to the underside of the plane at some point F_2. F_1 and F_2 will turn out to be the foci of the ellipse. This property, only a by-product of the proof, is by itself worth the price of admission.

Take any point P on the oval and connect it with V. VP cuts C_1 and C_2 at Q_1 and Q_2, and the length Q_1Q_2 is evidently constant for any position of P. But tangents from an external point to a sphere are equal. Hence

$$PF_1 = PQ_1,$$

also: $$PF_2 = PQ_2.$$

Adding: $PF_1 + PF_2 = PQ_1 + PQ_2 = Q_1Q_2 =$ *constant* for all P,

exactly the definition of an ellipse with foci at F_1 and F_2.

• • •

A ball resting on a desk, illuminated by a single electric light, casts an elliptical shadow on the desk, and the ball touches the

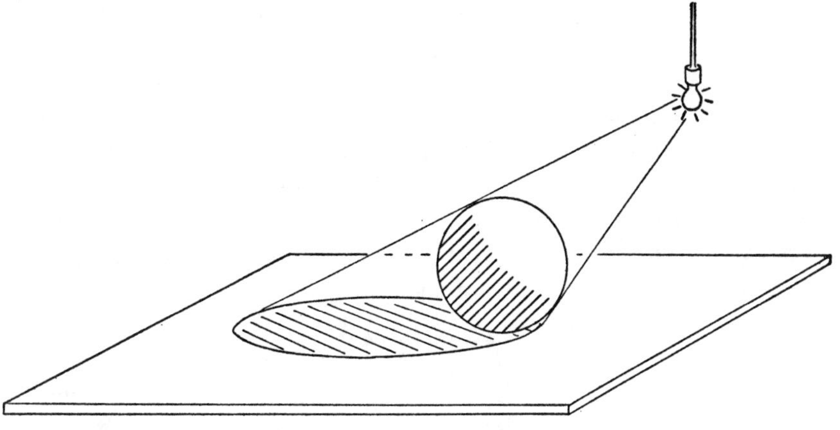

Figure 29.

desk at a focus of the ellipse. It is obvious why this is so if one thinks of the light-source as the vertex of a cone and the desk as the cutting plane. The ball is then the lesser Dandelin sphere. Suppose now that the ellipse of Figure 29 is traced out in pencil on the desk surface, and then the ball and the lamp are removed. To an observer who places his eye where the electric light was, the ellipse looks like a circle. Are there any other points from which the ellipse looks like a circle, and if so, how are they located? There are: a slightly smaller ball resting on the same spot may be made to cast the same shadow by a properly

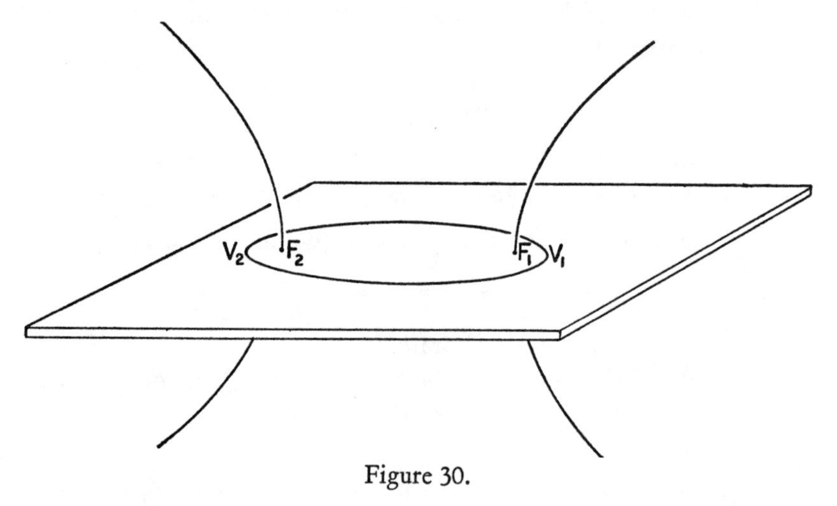

Figure 30.

placed light (nearer, of course). The 'locus' of all possible locations of the light (for suitably chosen balls)—in other words, the locus of all points from which the ellipse looks like a circle—is a *hyperbola*, whose foci are the vertices of the ellipse and whose vertices are the foci of the ellipse. The plane of this hyperbola is at right angles to the plane of the ellipse (Fig. 30).

• • •

You see ellipses every day, although your mind translates the visual images back to circles for you: whenever you look at wheels from an angle, for instance. But there is another common phenomenon which displays an ellipse without the need of any perspective or projection. Every time you tilt an ordinary tumbler of water the surface of the water, remaining horizontal, takes the shape of an ellipse (Fig. 31). Or you may use an even more realistic cone in the form of a paper cup. Nearly everyone has had a drink at one time or another from a conical paper cup and hence held in his hand a practically perfect ellipse; yet it is a curious fact that if you ask anyone for an everyday illustration of a conic section, you will always draw a blank.

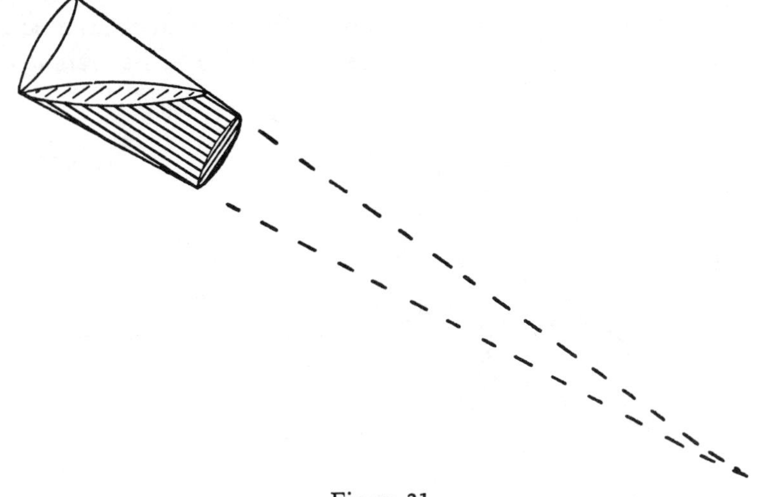

Figure 31.

The annoying tendency of the eye to see something without properly reporting it to the brain is not restricted to mathematical objects. A characteristic of the human mind seems to be that it sees only what it is looking for, which is by no means all of what is there. In fact, psychologists record innumerable experiments in which the subject thinks he sees something that is not there at all: his mind merely puts it there by association, so that afterward he is 'sure' that it was there because 'it must have been.' We had an illustration of this type of illusion in the dovetailed cube at the end of the previous chapter.

At Stonehenge, the famous prehistoric monument in England, there have recently been discovered axhead designs carved on some of the great ancient stones. 'The first ax figure was noticed accidentally in July 1953 by [R. J. C.] Atkinson while he was photographing a 17th-century inscription. Thereafter, he and others recognized more than 45 others. Once pointed out, most figures are unmistakable. Yet, although hundreds of antiquaries since 1666 have minutely scrutinized the stones and tourists have scratched their names thereon, no one before saw a single ax!'

• • •

The conic sections turn up repeatedly in mathematics and physics. The main cable of a suspension bridge, if its weight is

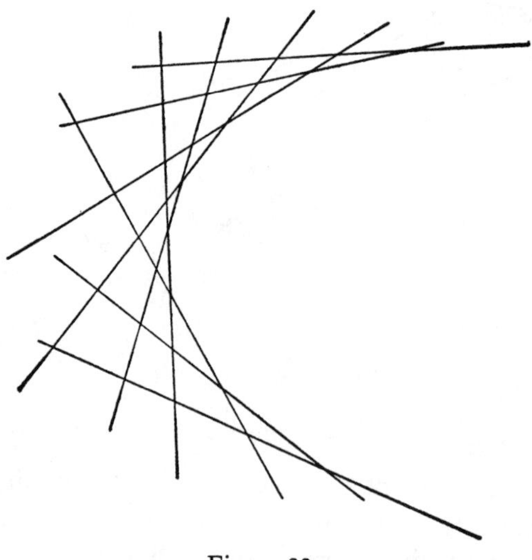

Figure 32.

small compared with the weight of the roadway, hangs in approximately a parabola. The surface of the liquid in a rotating bucket, which builds up around the sides because of the so-called centrifugal force, takes the form of a paraboloid of revolution. The 'inversely proportional' curves of the gas laws of physics are hyperbolic arcs. The orbits of planets and comets in the solar system are essentially conic sections. The list goes on and on.

The tangent lines to a curve, taken as an infinite family, are said to *envelop* the curve, or form its *envelope*. Some of the straight lines enveloping a parabola are shown in Figure 32. Observe that the parabola itself is not drawn; but it is easy to see where it lies.

Take a sheet of ordinary typewriter paper and crease it by laying one corner exactly on the opposite long side and making a

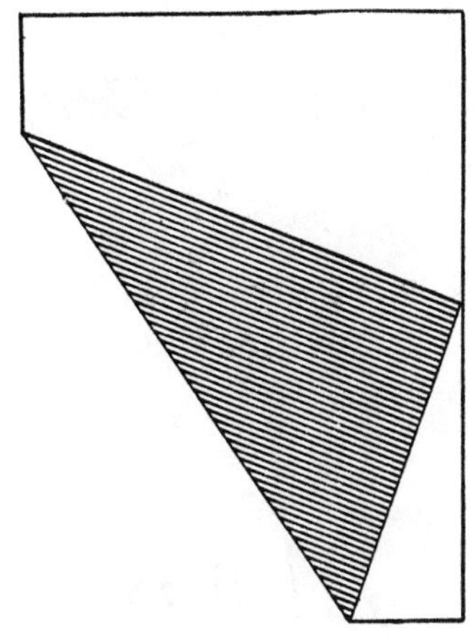

Figure 33.

fold (Fig. 33). Unfold, and crease it again, placing the same corner this time on a different point of the same long side. Repeat the operation a number of times. You will find that the creases envelop part of a parabola.

7 / *Wheels within Wheels*

Mark a point on the rim of a wheel. What is the path followed by this point as the wheel rolls along a straight line on level ground? Obviously the point must move somehow up and down in addition to moving forward; so perhaps you think the point follows a wavy line. But actually the path looks like the curve of Figure 34. It is called a *cycloid*.

The wheel is assumed to roll without skidding. But the ground doesn't move. This means that there is always one point of the wheel (the point of contact) that is not in motion. Thus the marked point whose career we are following comes to rest, though only instantaneously, once every revolution of the wheel. This implies that the motion down to and up from the contact position is all vertical: there is no horizontal component of motion at the cusps (points) of the curve. This is not the behavior that most people predict unless they give the question considerable thought.

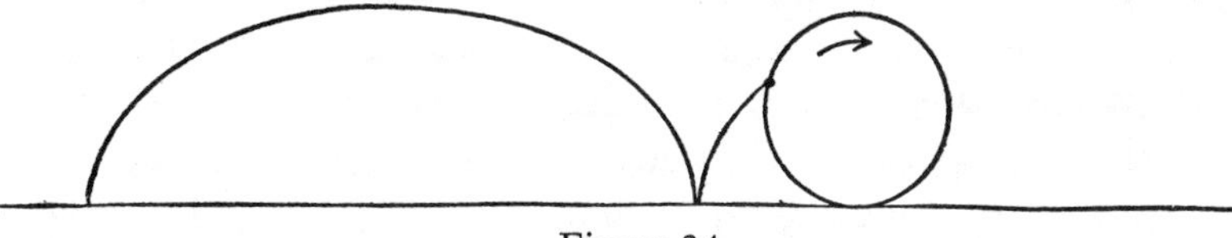

Figure 34.

But it is easily verified by the calculus; the slope of a curve at any point is one of the easiest things the calculus gives us.

The cycloid has been the subject of much study, for it has very many interesting properties.

The length of one arch is exactly four times the diameter of the generating circle (the rolling wheel); and the area under an arch is exactly three times the area of the circle. It is surely astonishing that these ratios should both be so simple, when the curve itself is utterly non-circular.

Suppose now that we invert the cycloid curve (Fig. 35), and

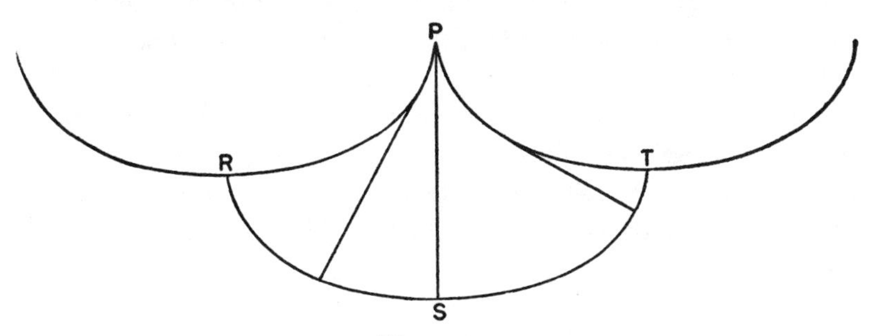

Figure 35.

imagine a string stretched around it from *P* to *R*, of length exactly twice the diameter of the generating circle. If we now unwind this string, keeping it taut, the point *R* traces out the curve *RST*. A curve generated in this way is called an *involute*. The involute of a cycloidal semi-arch is another cycloidal semi-arch the same size as the original.

A simple pendulum swings with a period approximately independent of the amplitude. To regulate a grandfather clock, you do not change the amount which the pendulum swings to the right and left: if you did, it would not affect the rate. Instead, you adjust the *length* of the pendulum. The question naturally arises, is there any kind of pendulum whose period is not only approximately but *exactly* the same for all amplitudes? The answer is yes: if the pendulum is hung from point P (Fig. 35) on a flexible wire PS which is constrained between cycloidal drums PR and PT, so that the bob follows the cycloidal path RST, then the time is the same for a small swing as for a wide one.

Suppose two small particles or pellets are placed in a smooth bowl and released at the same moment. Presumably, if they start from points at different heights on the side of the bowl, one will reach the bottom of the bowl before the other one. If friction is assumed to be negligible, what is the shape of the bowl in which the two particles will arrive exactly at the bottom at the same instant, from any two starting points whatever in the bowl? We have just answered this question in the last paragraph. The pendulum property says that the bowl should be cycloidal: the cycloid is therefore the *tautochrone*, or curve of equal descent.

Another interesting mechanical problem is this: along what path from a point A to a lower point B should a frictionless particle slide under the influence of gravity alone in order to move from A to B in the *least* possible time? You have guessed it: along a cycloid. Thus this curve is also the *brachistochrone*, or curve of quickest descent.

• • •

As a circle rolls along a straight line, a point on its circumference traces out a cycloid. We ask now the converse question: along what curve must a circle be rolled in order that a point on its circumference shall trace out a straight line? One's first guess might be an inverted cycloid. But this would be very wrong; for if the length of the diameter is d, the length of the cycloid arch is $4d$, whereas the length of the circumference is only πd. Thus after one revolution the circle will not have rolled to the end of one arch. On the other hand, any shorter cycloid would be too shallow; so we must abandon the cycloid.

We need a curve that dips to d units below its starting point, and whose length is equal to πd. A semicircle of radius d does just that. That the semicircle is indeed the curve we want is proved in the appendix: if a circle of radius r rolls without slipping around the inside of another circle of radius $2r$, a point (in fact *each* point) on the small circle 'rides back and forth' along a diameter of the large circle (Fig. 36). For various positions of C, arc SC = arc PC, and P moves from S to T.

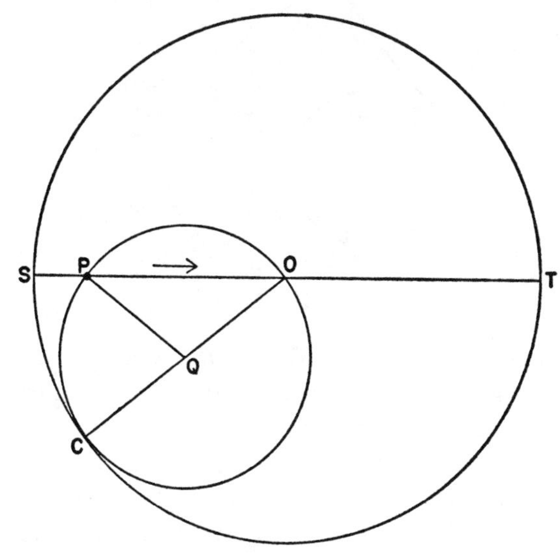

Figure 36.

Suppose next that the radius r of the small circle is not exactly half the radius R of the large circle, but some other fraction. The resulting curve traced out by a point P on the circumference of the small circle is called a *hypocycloid*. P returns to its starting point eventually if and only if R/r is *rational;* it returns every time around if and only if R/r is *integral*. If $R/r = 3$, we have the three-cusped hypocycloid (Fig. 37) sometimes called the *deltoid*.

• • •

The Greek mathematicians concerned themselves with geometrical constructions that were possible by using only the com-

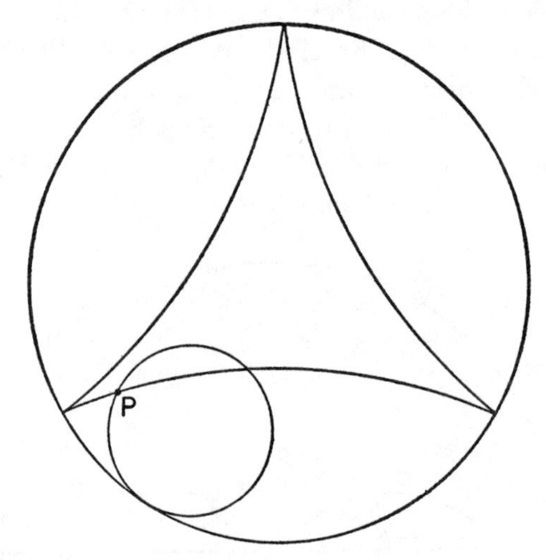

Figure 37. Deltoid.

pass and (unmarked) straightedge. The question had long been asked, what regular polygons are constructible with straightedge and compass alone? It is an easy matter to inscribe a square in a circle, and hence to obtain any polygon of 2^n sides. Likewise a hexagon is easy; and hence a regular polygon of 12, 24, $\cdots$ sides. Also the ancients knew how to construct a pentagon. But is a regular polygon of 7 sides constructible, or 9, or 11, or any other with an odd number of sides? They did not know.

It remained for the young Gauss to give the answer to this historic question. He discovered it at the age of 17, and was so intrigued by it that he decided then and there to make mathematics his career—luckily for mathematics. Gauss's discovery was that a straightedge-and-compass construction of a regular polygon having an odd number of sides is possible if and only if that number is a *prime* Fermat number, or a product of different prime Fermat numbers. A Fermat number, you will recall from Chapter 2, is a number of the form $2^{2^n} + 1$. Since there are only five known prime ones, the number of regular polygons of an odd number of sides constructible by straightedge and compass is relatively small. After the pentagon, the next is the 15-gon, whose side is found by suitably combining the information available from the triangle and pentagon. The next is the 17-gon, which requires an independent construction; and the next the 51-gon (3×17).

It is a fairly safe guess that no one reading about Fermat numbers for the first time would ever dream that they had the remotest connection with the construction of regular polygons.

• • •

It is easy to give a *mechanical* construction of a regular n-gon of *any* desired number of sides. Given a circle of diameter d, it is a simple ruler-and-compass construction to divide d into n equal parts. Using one of these parts, of length d/n, as a diameter, draw a second circle tangent to the first, and mark a point P on its circumference. Roll the small circle around the large one, marking every point of contact of P on the large circle. Since there will be exactly n such points, all equally spaced around the large circle, they can be used as the vertices of the regular n-gon. Of course this is by no means a straightedge-and-compass construction. The Greeks did not allow rolling circles as drawing tools.

• • •

What is the least area in which a ship can be turned end for end?

To reduce the problem to more precise mathematical language, what is the shape of the boundary of the least possible area swept out by a line AB when it is moved in the plane in such a way as to reverse its direction? Certainly a circle of diameter AB will suffice as a possible area; one need only pivot the line through 180° about its center. But this is not the *least* area. It was long thought that the deltoid might be the least. A deltoid just big enough to turn AB around in has only half the area of the circle (Fig. 39).

Here is an amazing instance of the kind of thing that can happen in science. A simple problem was studied by generations of

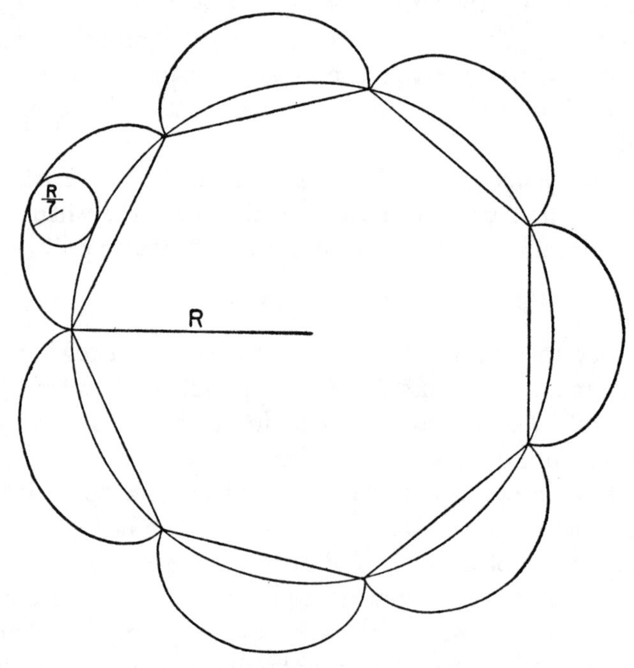

Figure 38. The epicycloid used to construct a regular 7-gon.

mathematicians, most of whom did new and important work on far more difficult topics; and yet not one of them had the insight, or inspiration, or whatever it takes to make original discoveries, to take the next step, which now seems perfectly natural and obvious, toward the solution of the problem. Of course, when once you have found out how to proceed, the way often seems 'obvious.' The finding out is the important part.

In this problem the next step after the deltoid is what for lack of a standard name we might call a 5-oid. It is pictured in Figure 40. Its area is substantially less than (about three-quarters of) the area of the deltoid for the same length *AB*. Well, then, let's try a 7-oid. Sure enough, the area is smaller still. So of course

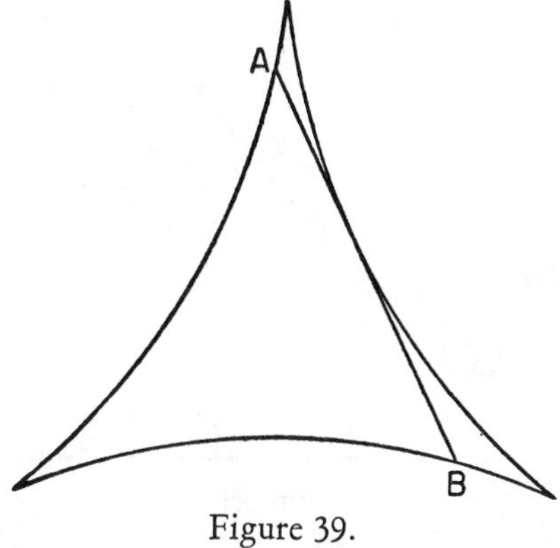

Figure 39.

we keep going; and A. S. Besicovitch, whose 1928 paper suggested this process, proved at the same time that *there is no least*

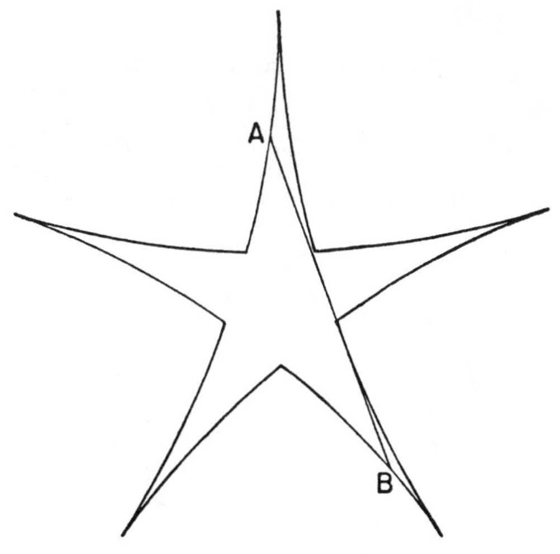

Figure 40.

area: the area may be made arbitrarily small by increasing the number (and length) of the spikes of the *n*-oid.

• • •

A circle is a *curve of constant width*. This means that the distance between parallel tangents is constant (Fig. 41). Are there any

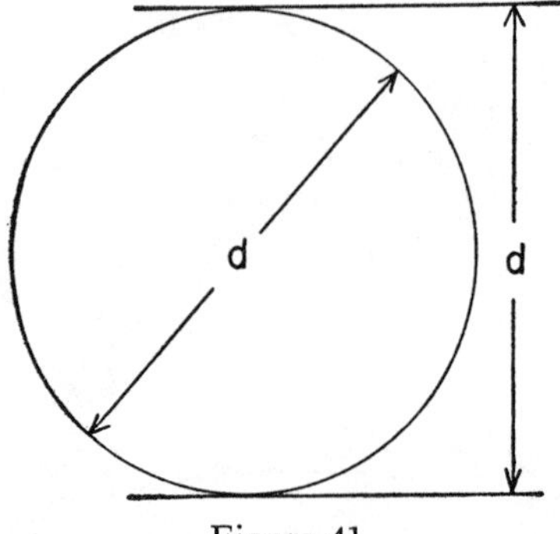

Figure 41.

other curves of constant width? Yes, indeed. Figure 42 shows one, formed by drawing equal arcs from three vertices of an equilateral triangle, each arc with radius equal to the side of the triangle.

Cylindrical pipes are often used as rollers to move crates or other heavy objects. A roller of the shape indicated in Figure 43 could be used just as well. Since its cross-section is of constant

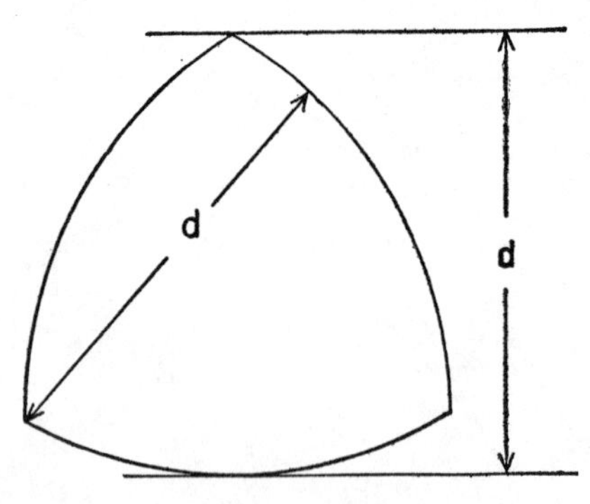

Figure 42.

width, the crate would roll along absolutely smoothly, without friction and without any up-and-down motion, despite the 120° corners on the roller. This is hard to believe even when you know it to be true; I doubt if you would have much success selling these rollers to the moving man.

• • •

A circle is usually defined as the path of a point which moves in a plane so that its distance from a fixed point is constant. But there is another quite different definition, credited to Apollonius: the path of a point which moves so that the ratio of its distances

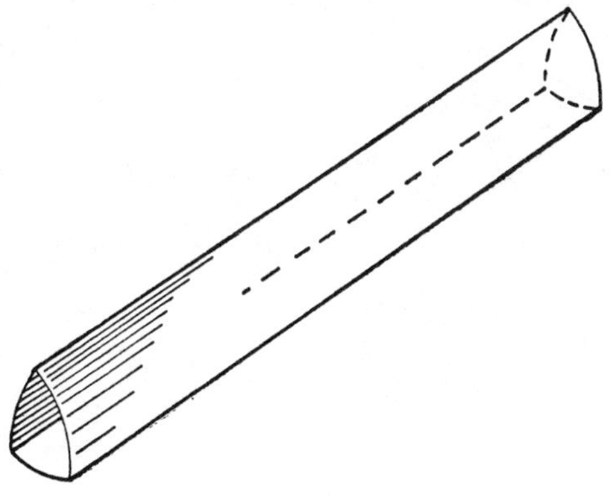

Figure 43.

from two fixed points is constant (Fig. 44). Note that the point R is not the center of the circle.

Suppose a ship leaves the point R and steams in a fixed direction at a constant speed. A second ship, leaving Q at the same time, can go twice as fast as the first ship. What direction should the fast ship take in order to intercept (catch) the slow ship as quickly as possible? The navigator of the fast ship should plot the Apollonius circle of the two points Q and R, and should then steer straight for the point P where the slow ship's course cuts the circle. Obviously the two ships will arrive at P simultaneously.

This solution is an approximation in that it assumes that the surface of the ocean is a plane, which it is not. If the earth's sphericity is taken into account, the problem becomes more complicated. I do not believe it has been solved.

• • •

A fast government patrol boat is overhauling a fishing trawler. Suddenly a fog descends, whereupon the trawler changes course but not speed, and proceeds straight in a new direction which is not known to the patrol boat. How should the patrol boat proceed in order surely to intercept the trawler in the fog? Yes, it can be done. (As before, assume a plane ocean.)

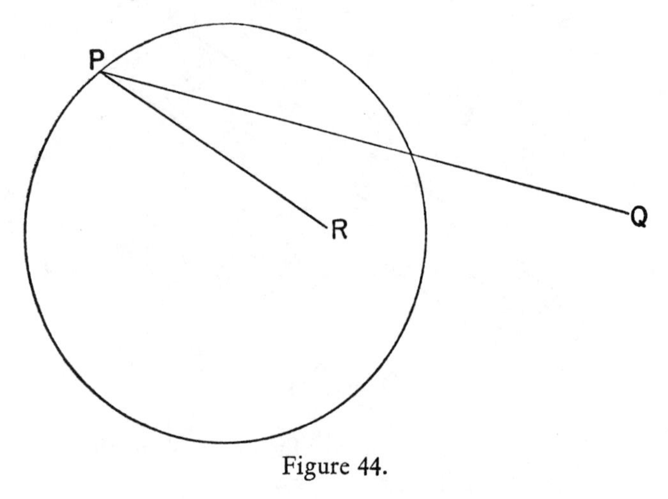

Figure 44.

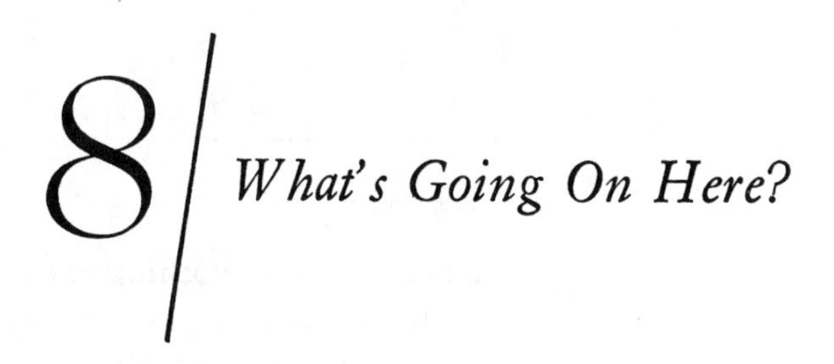

8 / *What's Going On Here?*

In this chapter we shall pause to do a little necessary stock-taking. But first an example.

On a piece of ruled paper, draw a pencil line diagonally from corner to corner (Fig. 45). Now wrap the paper around a right circular cylinder whose circumference equals the length of the paper (Fig. 46). The pencil line still makes equal angles with all the vertical lines, but it now has the form of a screw curve, or *helix*.

What is the analogous curve on a sphere? For instance, what would be the course of an airplane flying so that it crosses all the meridians of longitude at a constant angle? It is called a *loxodrome*, the course of an airplane whose compass heading does not change. One circuit is shown in Figure 47. The airplane must eventually fly round and round either the north or the south pole in a tightening spiral. Note that theoretically the pole is never reached: the number of 360° turns in a loxodrome is infinite. Yet

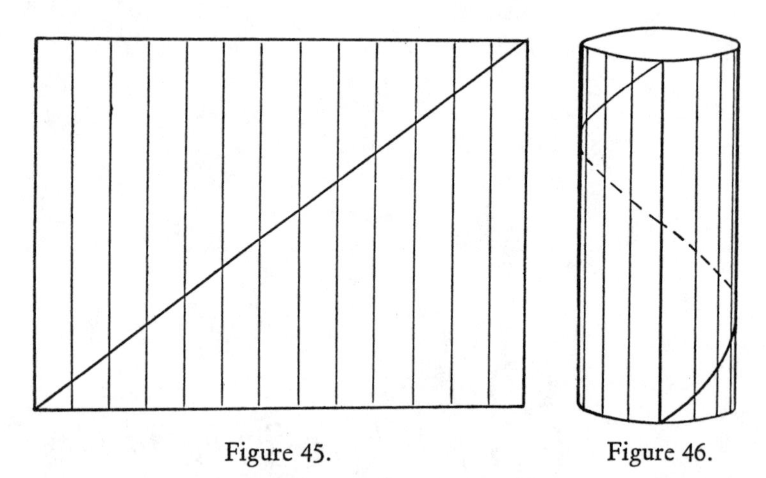

Figure 45. Figure 46.

the total length is finite. There is nothing extraordinary about this; it simply means that the lengths of the succeeding circuits form a convergent series. The total length of a loxodrome that goes from one pole to the other cutting the meridians at a constant angle α is the length of one meridian multiplied by sec α (Fig. 48). This result is suggested by observing that it is approximately true 'piecewise' for small sections of the curve. In the small, $ds = dy \sec \alpha$, if all curvatures are neglected. Essentially what the calculus does is integrate (sum) an infinite number of infinitely small elements of this type. Since sec α is a constant, it can be factored out of the summation; the dy's add up to the length of one meridian, and the ds's add up to the length of the loxodrome.

• • •

If you got lost somewhere in the middle of the last paragraph, take heart: it is no more than you could have expected. The example tries to give you a short course in the integral calculus and an introduction to differential geometry in half a page, which simply cannot be done. But it is an illustration of the ease with which the calculus handles geometric problems of an otherwise

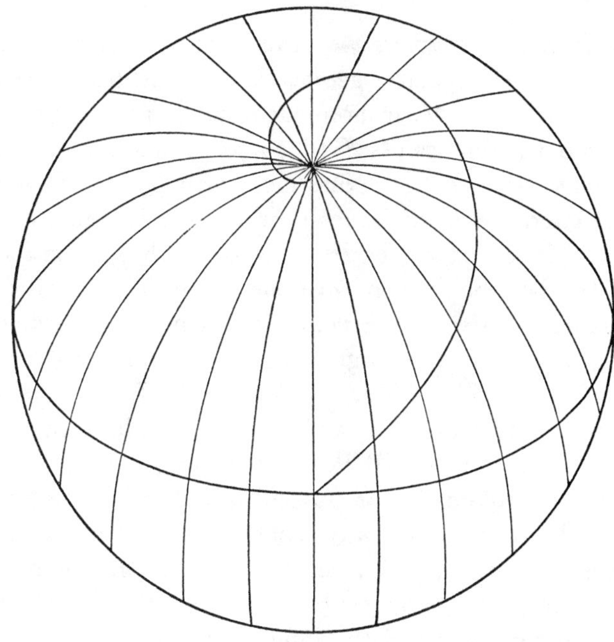

Figure 47.

almost insurmountable difficulty. Calculus is the beginning of the great branch of mathematics known as analysis, the main stream of development in the subject since the seventeenth century.

Many properties of the curves mentioned in the last chapter are easily derived by calculus methods. The Greeks had no calculus, or even any analytic (co-ordinate) geometry; and they consequently were forced to resort to ingenious but usually laborious methods to prove their theorems 'synthetically.' Today geometry and calculus are closely interwoven. Many otherwise tough geometric problems are solved by the methods of analysis. Indeed, modern analysis cannot be excluded completely from any

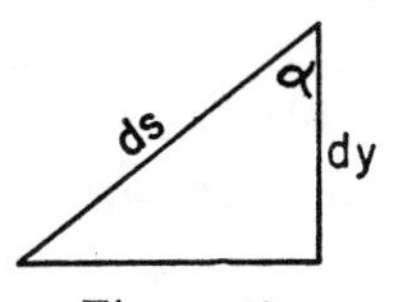

Figure 48.

other branch of mathematics. Even in number theory, the last stronghold of the *discontinuous*, the methods of analysis, where *continuity* is the password, are yielding more and more fruitful results.

In the foregoing chapters we have tried to stick for the most part to the simplest possible geometric methods, based frequently on visual evidence. Most non-mathematicians are willing to accept a proof if they can 'see' that it is so in a diagram. Therefore we have tried to help you along with plenty of diagrams. It might interest you to know that ultimately this method is worthless. Modern mathematical treatises eschew diagrams, and some of the best books contain none at all. This is all part of their insistence on *rigor*, the watchdog of modern mathematics. Our reasoning about the length of the loxodrome was about as non-rigorous as it could be. We tossed differential elements around with the reckless abandon of the eighteenth century, when the calculus was still in its struggling youth. Mathematicians had to pay dearly for the enthusiasm that tempted them to charge ahead without always consolidating their gains. They knew the calculus 'worked,' in the sense of giving the right answers to practical problems; but they were stymied when they tried to make sense out of the infinitesimal which they used so freely. The harder they tried to explain it, the less convincing they sounded—possibly because they were scarcely convinced themselves. The result was an early antagonism to the work of Newton and Leibniz which gradually subsided but which did not finally die out until the infinitesimal was exorcised during the nineteenth and twentieth centuries.

We make free use of diagrams in this book. You will have to take our word for it that we do so in order to help you see things that are valid, and not in order to pull the wool over your eyes in an attempt to prove things that aren't valid. Mathematics without diagrams jumps at once to a level too abstract for our purposes.

The real difficulties with diagrams are subtle, and began to cause trouble only after mathematics had graduated out of the elementary stages. But then the troubles became in some cases quite serious, and were cleared up only by a careful and thorough housecleaning, including a re-evaluation of basic assumptions which had previously been glossed over as 'obvious.' One of the major accomplishments of modern analysis is that it has succeeded in setting up solid foundations on which all of mathematics can safely build.

PROBLEM: To construct an equilateral triangle on a given line-segment AB.

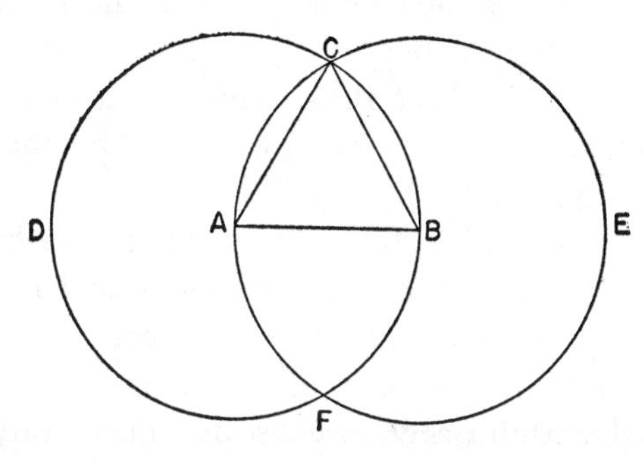

Statements	*Reasons*
1. With center A and radius AB, describe the circle $BCDF$. With center B and radius BA, describe the circle $ACEF$.	1. Postulate 3: A circle may be drawn with given center and given radius.
2. From the point C, where the two circles intersect, to the points A and B, draw the straight lines CA and CB.	2. Postulate 1: Through any two points one and only one straight line may be drawn.
3. $AC = AB$; also $CB = AB$.	3. Definition of a circle: a plane figure all of whose points are equidistant from a fixed point called the center.
4. $AC = CB = AB$, making triangle ABC equilateral.	4. Axiom 1: All things equal to the same thing are equal to each other.

There are many familiar optical illusions that serve to make us wary of believing too readily what our eyes tell us. But there are *mental* optical illusions that are much more difficult to detect. A classic example occurs in Proposition 1, Book 1, of Euclid's *Elements*. Euclid was acutely conscious of mathematical rigor, much more so than anyone else of his day or many a later day. His whole object was to set up a self-contained mathematical system, whereby every proposition followed from the application of a previously proved proposition or a previously stated axiom. Despite this commendable approach, he sometimes drew conclusions—probably unconsciously—from the evidence of diagrams. With the strong hint that *nothing* must be concluded on the basis of diagrammatic evidence, can you find the flaw in the Euclidean proof on page 89?

The difficulty is in step 2. It is not, as you might first suppose, that the two circles intersect at two points, C and F. The flaw is rather the assumption that the circles intersect at all. This is *not* evident, nor does it always happen. Two circles in the plane can lie wholly outside each other, or one may lie wholly inside the other. It is the way they are situated that makes them intersect. You say, 'But I can tell by looking at them that they must intersect.' This is what Euclid did, and that evidence is not admissible in this court. If we set out to prove something in mathematics, we must prove it. We are not allowed to say, 'Well, it's so because I can see it in the diagram.'

• • •

Consider the function $y = \sqrt[x]{x}$, read 'the x'th root of x.' Suppose we make a sketch of it on the usual Cartesian axes. In the positive range of x, things go along all right. If $x = 1$, $y = 1$; $x = 2$, $y = \sqrt{2}$; $x = 3$, $y = \sqrt[3]{3}$; etc. Each of these has a decimal value, and hence the curve can be plotted. It is continuous between the points where x has integer values, for a positive fractional exponent of a positive number always exists.

Putting aside the ticklish question of what happens when

$x = 0$, let us tabulate some values of y for negative x:

$$\text{If } x = -1, \quad y = -1, \qquad \text{real}$$

$$\text{If } x = -2, \quad y = \frac{1}{\sqrt{-2}}, \qquad \text{imaginary}$$

$$\text{If } x = -3, \quad y = \frac{1}{\sqrt[3]{-3}}, \qquad \text{real}$$

$$\text{If } x = -4, \quad y = \frac{1}{\sqrt[4]{-4}}, \qquad \text{imaginary}$$

This is pretty peculiar behavior. Furthermore, even the real part of the curve breaks itself into two pieces. Those interested in the algebra can investigate what happens when x is any negative fraction, $-p/q$. It turns out that if q is even, we have y positive real. But if q is odd, we have y negative real some of the time (when p is also odd) and imaginary the rest of the time (when p is even). Now we agreed in Chapter 5 that there are infinitely many rational numbers, p/q, and that they come as closely packed together as you please. Even the set of rationals p/q with p never odd is everywhere dense. Thus the points of the upper branch of the curve to the left (Fig. 49) appear to form a continuous curve. So, for that matter, do the points of the lower branch. Yet we know that there is no point on *either* branch corresponding to $x = -2$. Therefore, if we define the intersection of two lines as a point common to both lines, we see that the vertical straight line $x = -2$ crosses both branches of the curve without intersecting either of them!

The circles in Euclid's first proposition do not behave this way. But something must be done to guarantee that they don't. A postulate of continuity and its implications must be invoked to make the proof airtight.

Some mathematicians are so allergic to diagrams that they consider them ineligible to appear in any serious work. Others take the more moderate view that diagrams may be used to *illustrate* but not to *prove* anything. One must be quite certain

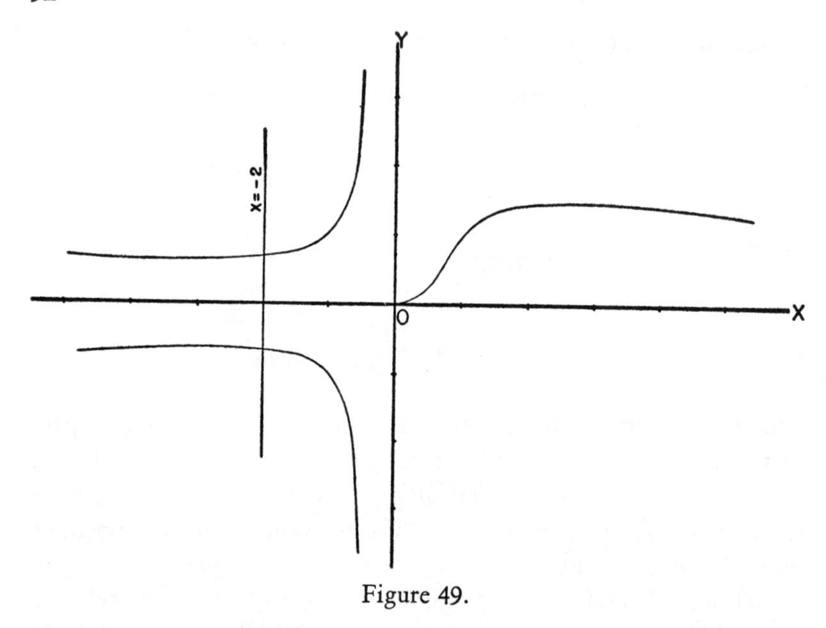

Figure 49.

that one in no way relies on them for information not furnished elsewhere in the proof.

• • •

The most elusive creature in the mathematical menagerie, the most difficult to tame and bring under the control of rigorous treatment, has been the concept of the infinite. Many mathematicians have wished to deny its existence; but it keeps reappearing, however hard we try to banish it. Georg Cantor (1845–1918) made great strides with his theory of infinite classes or *aggregates*. Some mathematicians think that Cantor at last subdued the troublesome infinite. Others believe he only caught it by the tail, and that it is still flailing about, breaking up the mathematical furniture.

An introduction to the Cantor theory would require many pages. We cannot do better than to refer the reader to the first two chapters of *Mathematics and the Imagination*,* where the problem of the infinite is presented entertainingly yet carefully, in

* By Edward Kasner and James Newman (Simon & Schuster, 1940).

language understandable to anyone. We shall content ourselves here with one theorem, easily proved, which may serve as a final warning against believing too readily what we see or what so-called common sense tells us 'must' be so. The theorem says that a line one inch long contains exactly as many points as a line two inches long.

Perhaps you think we ought to begin the proof of this apparently preposterous theorem with a definition of a point. But it turns out, after a couple of millennia of unsatisfactory attempts, that a point is best left undefined. It is something like trying to define the number '1.' Everybody knows what 1 means. To define anything so basic is not only unnecessary but probably impossible. The same may be said for a point. Even though no one has ever seen a point (it has no size), everyone has a pretty fair idea of what the concept means—really a very complimentary commentary on the intellectual sophistication of modern man.

What *is* necessary to explain is what is meant by the phrase 'as many.' How many points are there on a line? Surely an infinite number, since between *any* two we can always place another. There are an infinity of points on a line one inch long and an infinity of points on a line two inches long. What right have we to claim that two apparently different-sized infinities are equal?

Cantor decided to extend the notion of 'the same number' from the finite to the infinite. If there are some people in a room, and some chairs; and if each person sits in one chair; and if there are then no people left standing, and no chairs unoccupied; then I think we would all agree that the number of people and the number of chairs in the room are the same. Note that we did not have to count the people or the chairs to arrive at this conclusion. We simply paired the people with the chairs in a *one-to-one correspondence*, and there was none left over. Cantor says that if the objects in one collection can be paired one-to-one with the objects in another collection in such a way that there is none left over from either collection, then the two collections contain the same number of objects.

What we have to show, then, is that *all* the points of a line one inch long can be paired one-to-one with *all* the points of a line two inches long, with none left over from either line. As usual, this can best be illustrated by a diagram. It can be readily validated analytically, however, by assigning co-ordinates to the points. If AB is one inch long and CD is two inches long, draw CAO and DBO (Fig. 50). Now any point P of AB can be matched

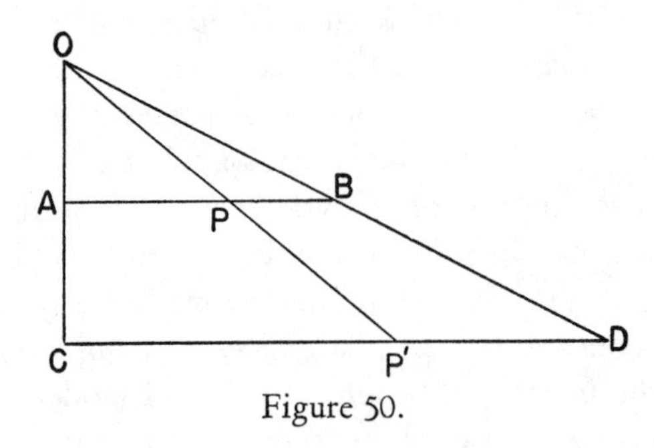

Figure 50.

with some point P' of CD by a straight line OPP'; and any point P' of CD can be matched with some point P of AB. An infinite 'fan' of straight lines through O will pick up exactly as many points from AB as it will from CD—namely, *all* of them. If you think some are 'left out' from CD, just connect them to O, and the connecting lines will pass through points of AB that have not yet been used. The correspondence really is one-to-one, which says that there are just as many points on AB as on CD, no more and no less.

By a slight extension of this idea it is shown (Fig. 51) that a line one inch long has the same number of points as a line of infinite length. Simply bend the one-inch line into two segments, and place the point O as shown.

This property that the whole is no greater than one of its parts is inherent in the arithmetic of infinite classes, which behaves not at all like ordinary arithmetic. We see now one more reason for not attempting to use the symbol ∞ in ordinary arithmetic and algebraic calculations.

A paraphrase of the paradox of Achilles and the tortoise (p.

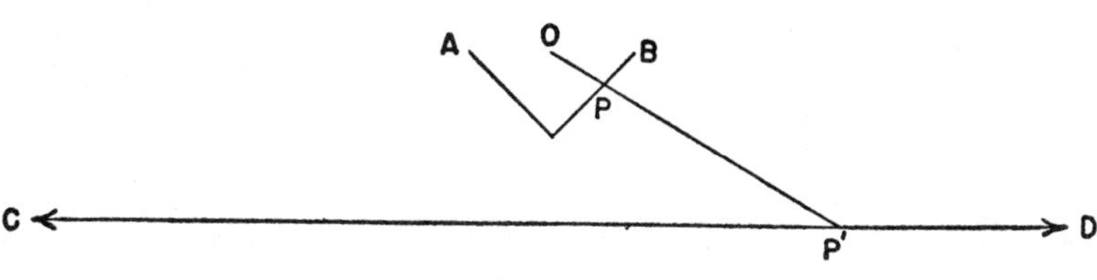

Figure 51.

20) due to Bertrand Russell can now be resolved: 'Achilles can never overtake the tortoise because Achilles has more positions to occupy on the way.' But positions are points; and we have just discovered that the long line traveled by Achilles does not contain any more points than the short line traveled by the tortoise.

• • •

The fact that a long line contains no more points than a short line is not a paradox. It leads to no self-contradictory results. There are, however, some real paradoxes in the theory of infinite classes, which have led to disagreements, some of which remain unsettled. One is equivalent to the problem of the barber who shaves all the men in the village except those who shave themselves. Does he shave himself? The answer is no, because he does not shave men who shave themselves. But if he does not shave himself, he is in the group that the barber shaves; and therefore the answer is yes. The resolution of this contradiction can lie only in the conclusion that there can be no such barber. The 'class of all classes not members of themselves' is in fact a non-existent class, although it was at one time an accepted member of the family.

9 / *All Shapes and Sizes*

One of the neatest accomplishments of elementary calculus is its ability to solve a wide range of maximum and minimum problems. With no trouble at all it delivers up the answers to questions such as those asked in Column A. Incidentally these can be very practical problems, often of commercial importance.

The differential calculus does not answer questions of the kind asked in Column B. These are of an essentially different nature, known as variational problems. Some (but not all) of them can be solved by the Calculus of Variations. Observe that in all the problems in Column A, only one dimension, or at most a ratio between two dimensions, is unknown. The shape is always specified. It is characteristic of the problems in Column B that the shape itself is unknown. The ordinary calculus can tell us only

A
(*Calculus Problems*)

1. A gutter is to be made from a long rectangular piece of copper by bending it as shown. For what angle θ will the area of cross-section (water carrying capacity) be the greatest?

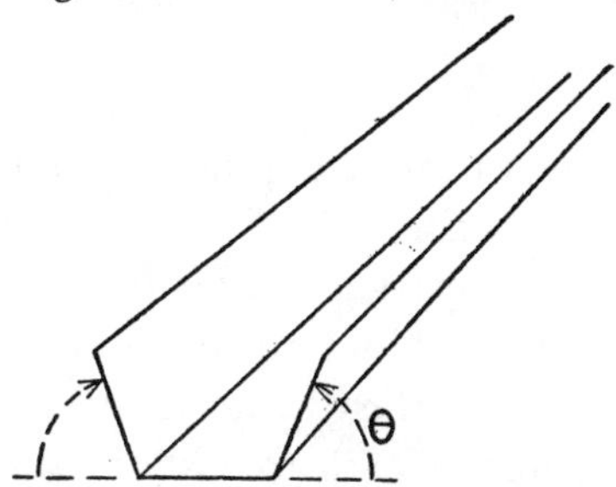

2. What point on a given plane curve is nearest a point not on the curve?

3. U.S. postal regulations state that the length plus the girth of a parcel-post package may not exceed 100 inches. Of all allowable rectangular parcels with square ends, which has the maximum volume?

4. If the quantity of metal used is the only consideration, what are the most economical proportions for a cylindrical quart can?

B
(*Variational Problems*)

1. To what form should the copper sheet be bent or curved for greatest area of cross-section? (Answer: a semi-circular gutter is the best. It carries more water than the gutter in Column A for any θ.)

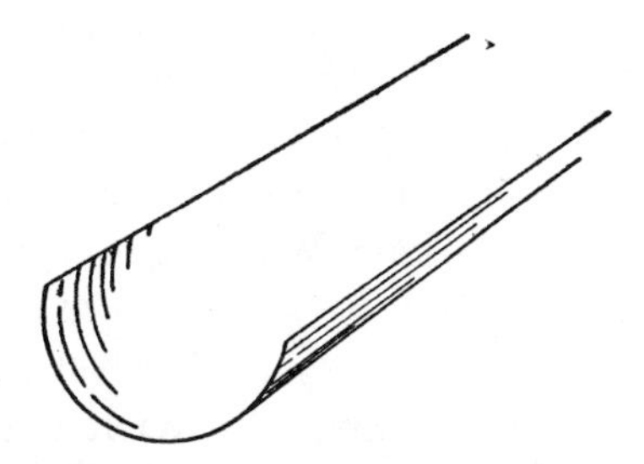

2. What kind of line is the shortest distance between two points? If this is too easy, do the problem on the surface of a sphere. If this is still too easy, some harder surfaces can be supplied at no extra charge.

3. Under the same restriction, what is the shape of the parcel of maximum volume? (Rectangular? Cylindrical? Spherical? Irregular?)

4. A man orders a silver napkin ring. He specifies the outside diameter and the width of the ring, and that the silver must be of a certain gauge (thickness); but he fails to mention the shape of the rest of the ring. Should the silversmith make it cylindrical, or will he save silver by using some other shape? If so, what shape?

what size; it requires more advanced methods to determine *what shape*. We have already met variational problems in this book. One is to determine the brachistochrone, the curve of quickest descent. By a method of his own invention Johann Bernoulli solved this problem in 1696. Hence this date perhaps marks the birth of the calculus of variations. But its growth has been slow and difficult, and much work remains to be done.

• • •

A necessary precaution in tackling variational problems is to make sure at the outset that a solution exists. This is often easier said than done. We have seen that mathematicians tried for years to determine the shape of the least area in which to turn a needle end for end, only to discover at last that there is no least area. A simpler example is: what is the shortest path that starts at A perpendicular to AB and ends at B? (Fig. 52.) Obviously some of

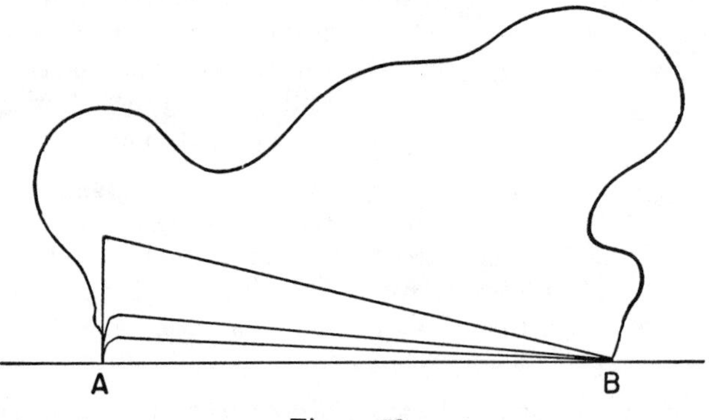

Figure 52.

the paths portrayed are shorter than others. Equally obviously there is no shortest. For we cannot use AB itself, inasmuch as it does not start perpendicular to AB; but no path perpendicular to AB can be the shortest, for we can always take another with a shorter perpendicular segment.

In this example the proof that there is no shortest path is immediate. In many variational problems, to decide whether the desired ***extremum***, as it is called, does or does not exist is just as difficult as to solve the problem. And in some cases (the needle problem, for example) the solution itself takes the form of a denial of any solution.

• • •

Consider a given number of points dispersed on the surface of a sphere, like observation posts on the earth. What is the best dispersal possible for 6 points? (Here 'best' means most widely separated each from each.) The answer is that they should be 90° apart: for instance, one at each pole and the other four at four equally spaced locations on the equator. These points are the 6 vertices of a regular inscribed octahedron.

What is the best dispersal for 5 points? Oddly enough, there is none. It would seem that one should be able to locate 5 points all farther apart from each other than 6, but it cannot be done. There is no single best pattern for 5: one at each pole and the other three anywhere along the equator, as long as they are not less than 90° apart, is still the best we can do. Not until we get down to 4 points can the dispersal be improved, and then the distribution is that of the vertices of the regular inscribed tetrahedron, about $109\frac{1}{2}°$ apart.

One might have been suspicious about 5 points, because there is no regular polyhedron with 5 vertices (see Fig. 17, Chapter 5). But as it happens, this criterion is spurious—there is a best dispersal for 7 points. And what is much more surprising, the best dispersal for 8 points is *not* the location of the 8 vertices of the inscribed cube.

The general problem of n points has not been completely solved. Although it seems a purely geometric question, it has applications in potential theory and electron mechanics.

• • •

Of all possible triangles having the same base and the same perimeter (distance around), which has the greatest area? This is one of the easiest of the isoperimetric problems, a class that contains some particularly sticky variational examples. We fall back on the conic sections for an elementary solution. It is clear that in the string construction of the ellipse, the string itself traces out all the possible triangles during its trip around the ellipse if the distance between the two tacks is considered as the fixed base. Since the area of a triangle equals one half the product of the base and the altitude, and since the longest altitude is the one through the center of the ellipse, we conclude that the maximum triangle is the isosceles one (Fig. 53).

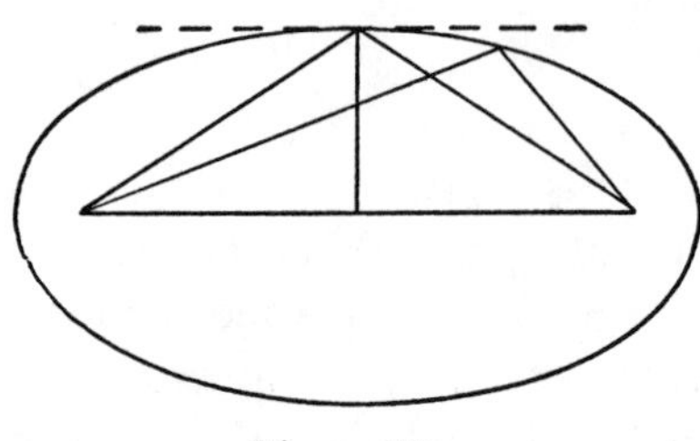

Figure 53.

• • •

What curve is assumed by a rope or chain suspended between two supports? (Fig. 54.) Does it hang in an arc of a circle, or part of a conic section, or possibly a cycloidal arch? None of these. The physical analysis produces a differential equation which integrates into a curve tailor-made to fit the hanging chain, called the *catenary*. The graph of a catenary *looks* very like a parabola, but it has an entirely different equation and different geometric properties. Figure 55 shows a portion of each. The equation of the parabola (broken line) is

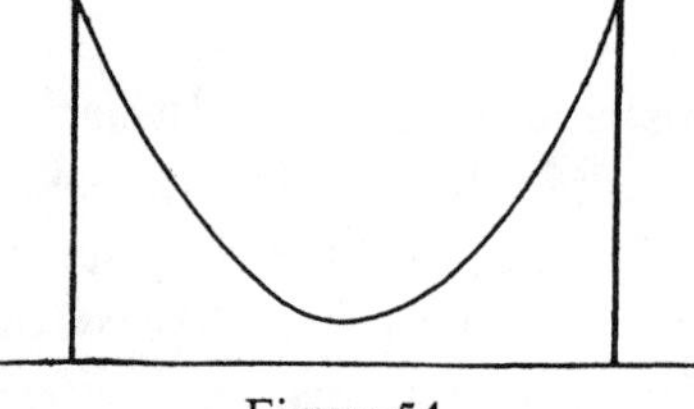

Figure 54.

$$y = x^2 + 1$$

and the equation of the catenary (solid line) is

$$y = \tfrac{1}{2}(e^x + e^{-x}).$$

The first is algebraic, the second transcendental. The catenary cannot be obtained by slicing a cone; neither will an unweighted chain hang in a parabola.

Here is a not very clever joke with which you may trap an unsuspecting victim. Suppose the rope of Figure 54 is attached

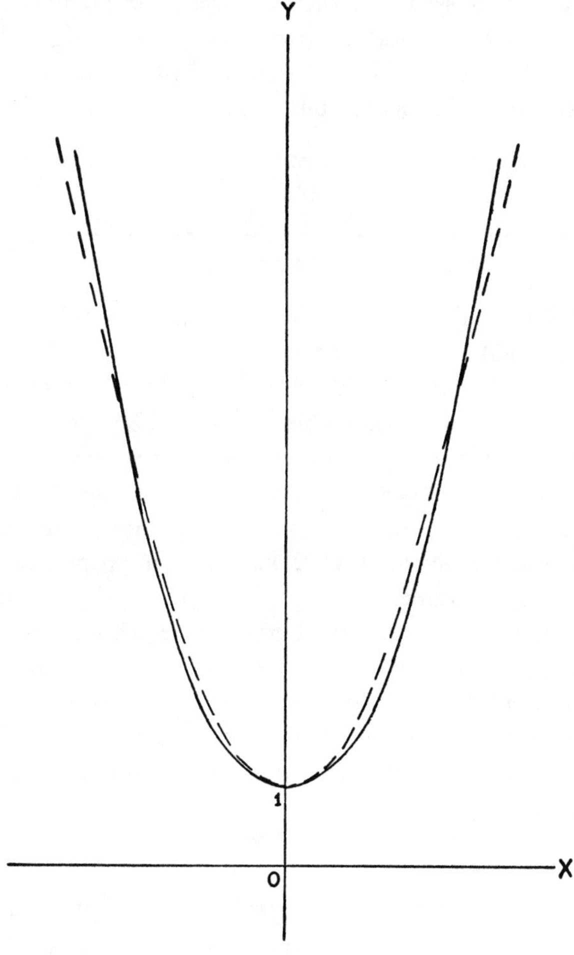

Figure 55.

to the tops of two posts, each 6 feet high; and suppose also that the rope hangs a little lower than in the diagram—so low that it just grazes the ground. If the rope is 12 feet long, how far apart are the posts?

• • •

A see-saw or teeterboard consists of a plank resting across a large log of semicircular cross-section (Fig. 56). If one end of the plank is depressed and then released, the plank will return to its original horizontal position of equilibrium. Is there any shape that the log can be given, instead of a semicircle, so that every position of the plank (until it slides off) is an equilibrium

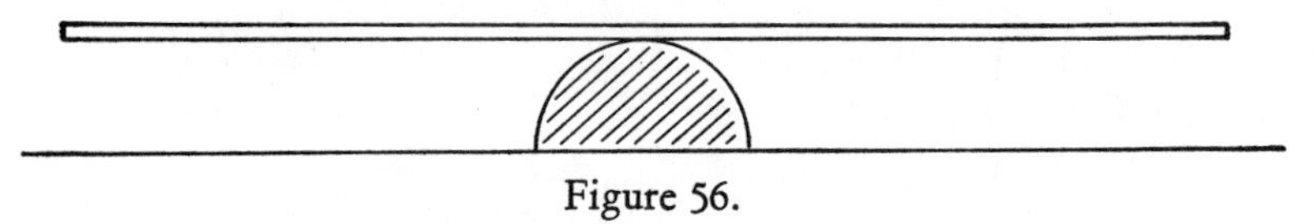

Figure 56.

position? Why, of course—an inverted catenary. On such a log the plank would stay tilted at any angle.

You may well ask how can there possibly be a connection between hanging chains and stable see-saws. The connection is indirect, but not far to seek. Starting with the physical condition that gravity is the only force acting on a hanging chain, one derives the equation of the catenary. This equation, considered now solely on its geometric merits, has certain properties. One of them is that its slope is proportional to arc-length measured from the center of symmetry. Furthermore, this property alone suffices to define the curve: no other curve possesses it. We then discover that this is just the property required by the see-saw if the center of gravity of the plank is to remain directly above the point of contact, which in turn is the requirement for stable equilibrium.

• • •

If a circular wire frame is dipped into a soapy solution and withdrawn, a disc of soap film remains stretched across the frame. This film will have the form of approximately a plane surface, sagging only slightly under the action of gravity. If a second wire circle is brought into contact with the first and the two are carefully separated, a film may be made to stretch between the two rings as in Figure 57. Because of a physical property of surface tension the film will automatically assume a *minimal sur-*

face; that is, it will answer physically the mathematical question of the least surface reaching between the two rings. This least surface is not a right circular cylinder. A surface like that of Figure 57 has considerably less area than the corresponding cylinder.

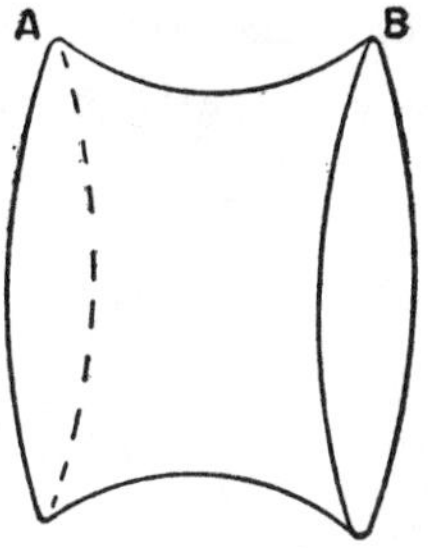

Figure 57.

• • •

We are now ready to return to the silversmith, whom we left at the bottom of Column B wondering how to make the napkin ring. His answer is given by the soap film of Figure 57: such a ring will require the least silver. But exactly what is the form of the curve from A to B? Yes, it's a catenary. Getting monotonous, isn't it?

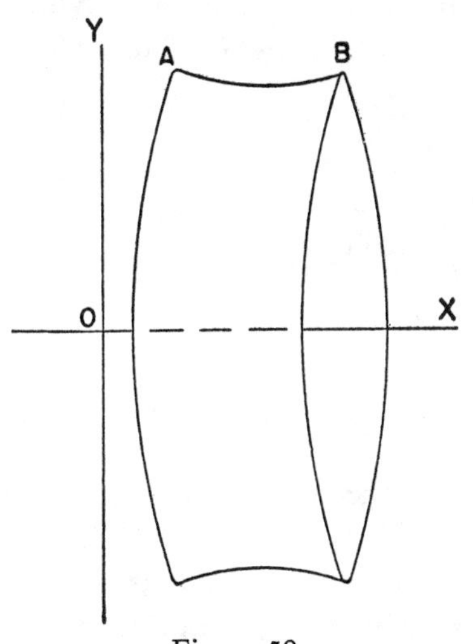

Figure 58.

Even an indication of the proof of this is beyond our scope here. It is a famous problem of the Calculus of Variations which was not completely solved until the present century. Its solution illustrates again the fact that you must be very sure that a minimum surface exists before you can claim to have found one.

We state the problem geometrically. Given two points, A and B, in the xy-plane (Fig. 58), what arc connecting them is the one which, when rotated about the line OX, generates the smallest surface area? There is an infinitude of different catenaries passing through A and B, but the right one can be found by the application of standard variational methods provided A and B are located more or less as shown in Figure 58. But what if the points are located as in Figure 59A, much closer to the axis of revolution? Now it is optically evident (and it can be proved) that the best solution would be to connect A to B by the bent line $ACDB$, and then rotate that. The area generated by the part CD is zero; and the area generated by the parts AC and BD are two small

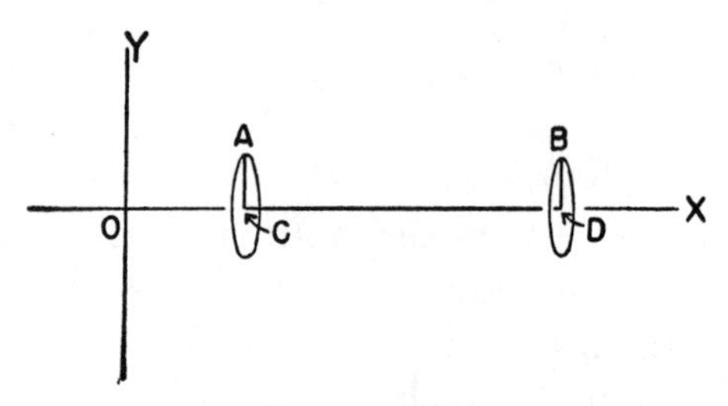

Figure 59A.

circles which together have less area than any curved surface made by an arc stretching from A to B. This is not a napkin ring; but there is no *least* surface with a hole in it through which the X-axis passes. For any curve of the kind shown in Figure 59B,

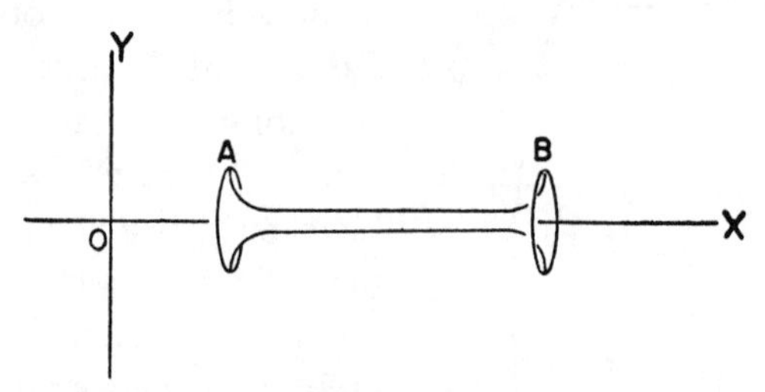

Figure 59B.

there is always another one closer to the limiting position and hence generating a smaller area.

The discontinuous solution $ACDB$ of Figure 59A was devised in 1831 by Benjamin Goldschmidt, who pointed out that this was the situation when the points A and B were close to the X-axis as compared with their distance from each other. *How* close, he did not say. The question is complicated and has no easily formulated answer. The complete solution was finally given in 1906 by H. F. MacNeish.

We have repeatedly watched different problems from widely diversified branches of science and mathematics lead ultimately back to a simple curve whose equation is an old friend. It is wrong, though, to think that this *always* happens. On the other side of the ledger, many hours of weary toil have been spent over equations that just aren't simple and cannot be made so. MacNeish's criterion is one of these. If A is a point on the Y-axis, then the solution of our minimal surface problem depends on whether B lies to the right or left of a certain curve. This is not an 'elementary' curve, and it can be plotted only by methods of approximation.

The whole problem is discussed in very readable fashion by G. A. Bliss.* That he devotes an entire 40-page chapter to it is an indication of the complexity of the problem.

• • •

We close this brief discussion of extremals with an application of the obliging tendency of soap film to form a minimal surface.

Suppose we make a clay pipe for blowing soap bubbles, with two bowls instead of one. These two identical bowls are connected by a 'Y' to one mouthpiece (Fig. 60). If the pipe is dipped in soapsuds and removed, and we blow into the mouthpiece, two small caps will appear and grow to the size of hemispheres, as shown in Figure 61 (A and B). Then if we blow a little more what happens? You may think that both bubbles continue to grow, as in Figure 61C. Actually something entirely different happens. All the facts are in your possession, and diagrams A and B are correct; only

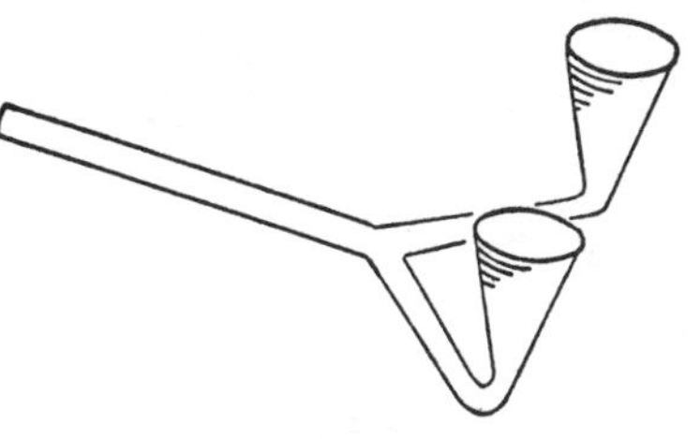

Figure 60.

* *Calculus of Variations*, No. 1 of the Carus Monographs (Open Court, 1925), a series published under the aegis of the Mathematical Association of America and dedicated 'especially to the wide circle of thoughtful people who, having a moderate acquaintance with elementary mathematics, wish to extend their knowledge without prolonged and critical study. . .'

Figure 61A.

Figure 61B.

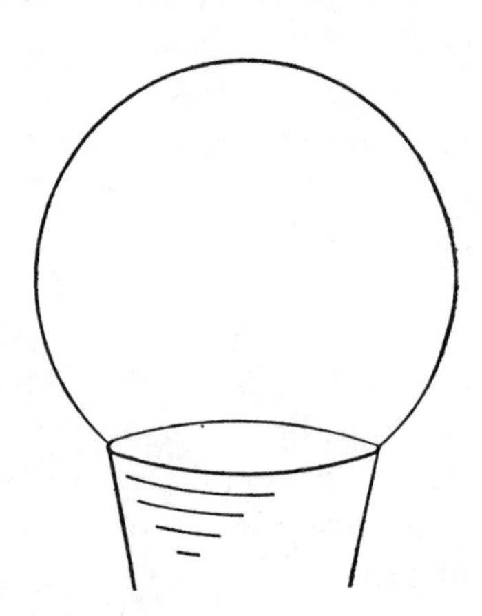

Figure 61C.

Enlarged views of twin pipe bowls, with bubbles.

C is incorrect. Can you think what the bubbles should look like? Or are your sentiments perhaps expressed in the following doggerel, written by a friend to whom I sent this puzzle?

I'd always thought that a soap bubble
Was a simple thing that caused no trouble
Until the day I saw one double;
   Then I went into a hurried huddle,
   My brain was in a worried muddle:
   'Twas like gazing into a muddied puddle.
   Large books into my arms I'd cuddle
   If they my head would un-befuddle;
   Alas, all I can say is, 'Wudd'le
I do with this double bubble trouble?'

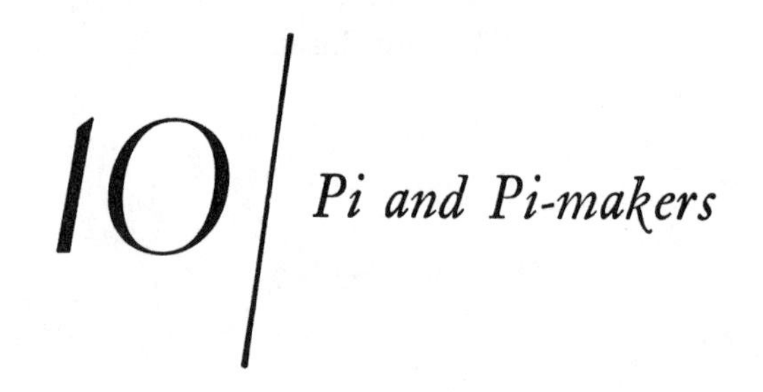

THE ALTERNATING SERIES

$$1 - \frac{1}{3} + \frac{1}{5} - \frac{1}{7} + \frac{1}{9} - \cdots$$

converges. Its sequence of partial sums, formed by cutting off the series first after one term, then after two terms, then three, etc., oscillates about a limit; but the swing above and below this limit (a certain definite number) becomes ever smaller as more terms are used.

The limit approached by these partial sums—in other words the limit of the series—multiplied by 4, equals π.

You are familiar with π as the ratio of the circumference of a circle to its diameter. You will also recall that this ratio 'does not come out even,' but that it is expressible, to 4 places of decimals, as 3.1416. How did scientists obtain this figure? Certainly not by measuring circles drawn on paper or the blackboard (which are not circles at all but only inaccurate attempts to *picture* circles). If you were to take a great many terms of the above series, add together the positive ones, subtract out the negative ones, and multiply the result by 4, you could arrive at 3.1416. It would be an arduous task, because this series converges slowly. But there exist series that converge much more rapidly, to simple expressions involving π; and it is by computing a number of terms of such series that π is actually calculated to any desired degree of accuracy.

The number π turns up in many different guises. We shall list a few of them. Many seem to have not the remotest connection with circles; and defining the ratio of circumference to diameter is not necessarily π's most important job, although certainly its oldest.

The following *infinite product* for π was devised by the Englishman John Wallis (1616–1703):

$$\pi = 2\left(\frac{2^2 \cdot 4^2 \cdot 6^2 \cdot 8^2 \cdots}{1^2 \cdot 3^2 \cdot 5^2 \cdot 7^2 \cdots}\right)$$

Vieta's product (1597), stemming directly from the areas of inscribed polygons of 4, 8, 16, 32, $\cdots$ sides, is

$$\pi = \cfrac{2}{\sqrt{\tfrac{1}{2}}\sqrt{\tfrac{1}{2}+\tfrac{1}{2}\sqrt{\tfrac{1}{2}}}\sqrt{\tfrac{1}{2}+\tfrac{1}{2}\sqrt{\tfrac{1}{2}+\tfrac{1}{2}\sqrt{\tfrac{1}{2}}}}\cdots}$$

We have also a continued fraction:

$$\pi = \cfrac{4}{1+\cfrac{1^2}{2+\cfrac{3^2}{2+\cfrac{5^2}{2+\cfrac{7^2}{2+\cfrac{9^2}{2+\cdots}}}}}}$$

We learn in the calculus that the area between a curve and the X-axis is given by the definite integral. The equation of the unit circle is

$$x^2 + y^2 = 1,$$

or

$$y = \pm\sqrt{1 - x^2}.$$

Its area is twice the area of the upper half, namely

$$A = 2\int_{-1}^{1} \sqrt{1 - x^2}\, dx = \pi(\text{square units}).$$

This integration requires a trigonometric substitution. The fact that it yields π depends upon the previous definition of π as the ratio of the circumference of a circle to its diameter. What the evaluation of this integral proves is that *if* π is defined as the ratio of circumference to diameter in any circle, *then* the area of the unit circle is π.

• • •

Imagine a very well-built oak floor whose boards, all exactly two inches wide, are fitted snugly together so that the very narrow cracks between the boards are essentially a set of parallel straight lines two inches apart. Suppose a needle two inches long is tossed or dropped on the floor. It will sometimes land across a crack and sometimes not. The *probability* that it will end up lying across a crack on any one fall can be shown to be $2/\pi$.

This unexpected hiding place of π was discovered by Count Buffon, the eighteenth century naturalist. It interested the mathematicians of the day because they thought that by tabulating the results of a large number of tosses one could obtain empirically the probability (see definition, beginning of Chapter 3); and then by equating this value to $2/\pi$ one could accurately evaluate π. The discrepancies of the experiment, however, owing to thickness of needle, width of cracks, etc., prevent extreme accuracy from being obtained by this method, even if one is patient enough to record carefully the results of thousands of tosses.

• • •

Suppose we took some specific measurement in a large sample, such as the height in inches of all male students in a university of 10,000, and tabulated the results. If we were to collect the individual measurements into groups that lay within, say, the same half-inch range, and make a *bar graph* of the numbers in each grouping, we should get a diagram that looked something like Figure 62. The highest bar might represent the number of

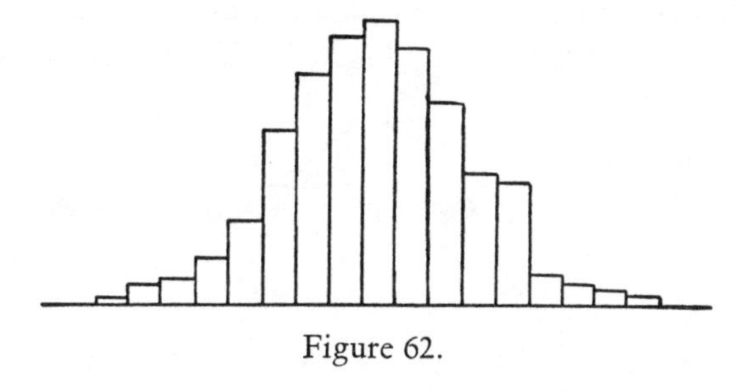

Figure 62.

individuals whose height was between $69\frac{1}{2}$ and 70 inches, and the frequencies would taper down on either side of this maximum number.

If now we could take an average of a large number of such

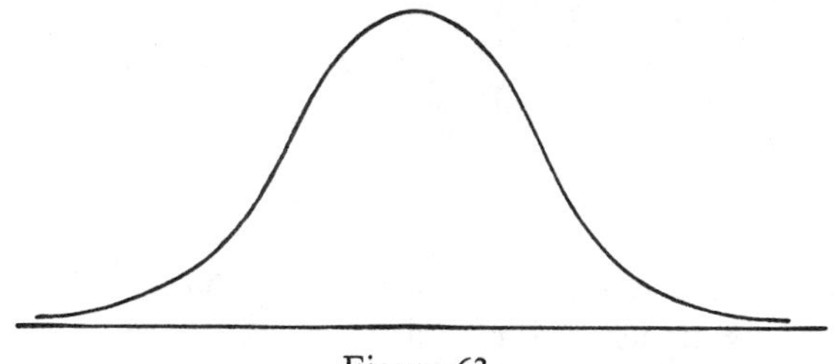

Figure 63.

samples, with more accurate measurements and smaller subdivisions, we should get a smoother graph. And it would tend to look more and more like Figure 63, where the area under the curve still represents the number of individuals in a sample.

This same frequency distribution occurs, theoretically, whenever measurements are taken from a sufficiently random sample. Examination grades; the lengths of the tails of a collection of adult white rats; the distances of bullet holes from the center of a target; a large number of careful measurements of the length of a physical object (such as a steel rod); all group themselves in this fashion. Even the frequency of heads obtained in a large number of tosses of a coin falls into the pattern. So we are back with our old friend, the Pascal Triangle. The binomial distribution approaches the frequency distribution as you go down the rows of the triangle. Look, for instance, at the coefficients of the 8 row plotted in this fashion (Fig. 64). If we took larger values

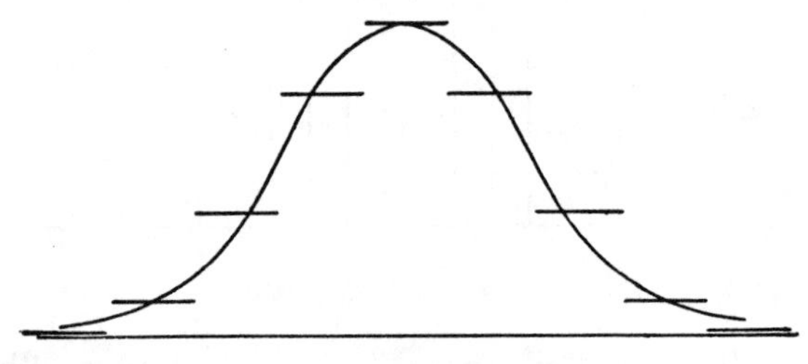

Figure 64.

of n in the expansion of $(a + b)^n$ and scaled the graph down accordingly, the coefficients would come ever closer to tracing out the dubbed-in curve of Figure 64.

The equation of this bell-shaped *normal distribution* curve, as it is called, is, in its simplest form,

$$y = e^{-x^2}.$$

It is essential to statisticians, who use the curve almost daily, to know the area under it for various values of x, since the area represents the frequency in a given range. The area from the midpoint to any positive x (Fig. 65) is represented by

$$A = \int_0^x e^{-x^2}\, dx.$$

But unfortunately not all integrals can be evaluated as easily as was the one for the area of the circle; and this happens to be one of the ones that simply cannot be carried out. The area is there:

but the statistician has to be content to refer to tables for its value, which can be obtained by methods of approximation, to any desired degree of accuracy.

The normal distribution curve approaches the X-axis asymptotically to the right and left, without ever quite reaching it. Yet the *total* area under the bell is finite. This means that however far out you go to measure the area, you never get an area greater than a certain limiting value. Even though the shaded

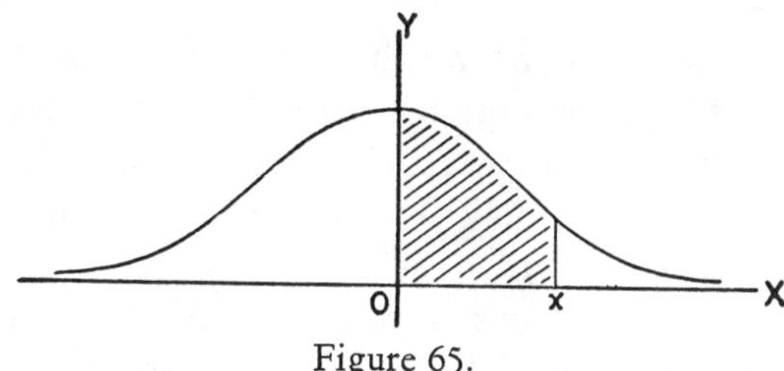

Figure 65.

area of Figure 65 cannot be evaluated directly, the *whole* area under the curve can be accurately determined, by a method which is one of the showpieces in the repertory of every teacher of advanced calculus. It works out to be *the square root of* π. This is a strange way to run across π again; but we have a still stranger one in store.

• • •

Logarithms of negative numbers don't exist, you were told in high school? If so, you were told wrong. The correct statement is that they are imaginary. There is a world of difference.

An imaginary number is a number involving the square root of a negative quantity, such as $\sqrt{-1}$. The term *imaginary* is a misnomer. The square root of minus one is no more imaginary than the number 2 is. What is so real about 2? Have you ever seen 2, or touched it? Of course not: it is a *symbol*. 'But,' you protest, 'I can see and touch 2 books; I cannot see or touch $\sqrt{-1}$ books, because there is no such thing as $\sqrt{-1}$ books.' There you are absolutely right. But all you have said is that 2 may be used for counting objects, while $\sqrt{-1}$ may not be used for counting objects. If that is your only standard, *minus two* is also 'imaginary,' because you cannot see or touch -2 books. Yet you are not nearly so upset about -2 as about $\sqrt{-1}$. The fact is that -2 *was* considered imaginary by ancient civilizations, and many centuries had to pass before it was accepted as a member of the family of numbers. So we find that it is only prejudice and lack of familiarity with $\sqrt{-1}$ that make people uneasy about its existence—a prejudice which anyone who works with it for a little while very quickly forgets.

We might remark in passing that $\sqrt{-1}$ is a most useful quantity in physics, engineering, and other aspects of 'real life.' The theory of alternating currents and the theory of airflow patterns, to mention but two widely separated topics, could scarcely be studied without the aid of $\sqrt{-1}$.

One of the most famous formulas in all mathematics states that π is the numerical value of the (imaginary) natural logarithm of -1. That is,

$$\pi = \frac{\log(-1)}{\sqrt{-1}}.$$

Written in exponential form, it looks like

$$e^{\pi\sqrt{-1}} = -1.$$

Not only is it an extraordinary place to find π; in addition, it provides an unexpectedly simple connection between π and e, two of the most important constants of mathematics.

• • •

Pi is an irrational number. We know, however, that we can find rational numbers as close as we please to any irrational. Three and one-seventh, sometimes used because it is compact and easy to handle, is not a very close approximation to π, but it is plenty good enough for many ordinary computations.

$$3\tfrac{1}{7} = 3.142857\cdots,$$

our old friend of Chapter 2. A much closer fraction is

$$3\tfrac{16}{113} = 3.1415929 \cdots$$

which compares with

$$\pi = 3.1415927 \cdots$$

• • •

In every book of mathematical tables or engineering handbook one finds a value of π to at least ten decimal places. This is more than good enough for all practical applications. If we knew the exact radius of the earth, and if the earth were an exact sphere, π to 10 decimal places would yield an error of not more than an eighth of an inch in the circumference.

Values of π have been calculated to very many more than 10 decimal places. A man named Shanks carried out the work to 707 decimal places (completed in 1873). Just why Shanks devoted many years to this job is hard to say. π had long been known to be irrational, so he could have had no hope of discovering a long periodic repeating decimal (see Chapter 2). He had previously announced a value of π to some 500 places; and perhaps he got so interested that he just couldn't stop, although one fails to see the fascination in such a project. It turned out that the last couple of hundred of Shanks's 707 places were wrong anyway, but nobody cared enough to find this out until 1946.

In 1940, Kasner and Newman wrote in *Mathematics and the Imagination:* 'Even today it would require 10 years of calculation to determine π to 1000 places' (p. 78). In 1940 there were plenty of fancy mechanical calculators to help do the work; but the big new *electronic* calculators had not yet hit the headlines. The huge superiority of these modern wonders in purely arithmetical calculations is strikingly exemplified by what happened when π was subjected to a concerted attack. Taking over the ENIAC machine during a long weekend (Labor Day, 1949), when its expensive services were otherwise unemployed, three mathematicians 'programmed' it to cook pi for a while, which it did with unprecedented speed and accuracy. At the end of 72 hours, π was displayed to a heroic 2000 decimal places, presumably all of them correct. May the soul of Mr. Shanks, who spent most of his life drudging out 707 places (some incorrect), rest in peace.*

• • •

The classical problem of 'squaring the circle' is the geometric equivalent to the problem of evaluating π. If π came out even, it would be possible to construct geometrically, with compass and straightedge as the only tools, a square whose area would be the same as the area of a given circle. It would even be possible if π did not come out even (rational), provided it were a number expressible with a finite number of square root signs. It was long suspected by mathematicians that it could not be so expressed, and the proof—a very difficult one—was finally developed in 1882 by Lindemann. Thus this age-old problem of geometry was settled in the negative by higher analysis: it is definitely not possible to construct with compass and straightedge a square whose area equals the area of a given circle.

There have been many attempts, most of them by amateurs sadly lacking in mathematical background, to 'square the circle.' You might suppose that the proof in 1882 of its impossibility would have made it clear that these attempts were a waste of time. But if that is what you think, you are not well acquainted with the strange company of circle-squarers. These individuals persist in claiming that they have 'proved' astonishing mathematical results. Usually their claims take the form of a new value for π—the true and only correct value, of course; and it is generally not especially close to 3.1416. Also, oddly enough, no two circle-squarers ever arrive at the same 'new' value.

* Since the initial publication of this book, π has been "evaluated" by far faster and more sophisticated computers—to what purpose it is hard to say. No pattern of the digits, even of the most involved kind, has ever been discovered. An essay by Richard Preston, "The Mountains of Pi," in the *New Yorker* of March 2, 1992, includes much data on all aspects of the subject. The calculation has now been run out to more than a *billion* decimal places. [1994 note.]

No professional circle-squarer, if we can so entitle these misguided gentlemen, has one-tenth enough mathematical background to understand Lindemann's proof or even the streamlined modern versions of it. If he did, he would never have taken up his hobby. The circle-squarer is in fact not in the least amenable to reason. You can knock him down in any number of ways, and he will bounce right back up again.

De Morgan devoted large portions of his *Budget of Paradoxes* to raking the circle-squarers of his time over the coals in grand style. But 1882 did not mark the end of this strange pursuit. Lest you think that circle-squaring went out with the nineteenth century, pause a moment and consider the following. They will strain your credulity, but they are facts.

On the shelves of many college libraries are copies of a book, privately printed (of course) in 1935, with the magnificent title, 'Behold! The Grand Problem No Longer Unsolved; The Circle Squared Beyond Refutation.' This book is on those shelves because, according to the author's statement in the introduction, he sent it at his own expense in order that the world might be enlightened to the truths he had discovered. Pi soon emerges from the ornate pages of this gem of literature as $3\frac{13}{81}$, which decimalizes to $3.1605 \cdots$. The author's supreme self-confidence is clearly indicated by the following circumstance. An immediate geometric consequence of his value of π is a right triangle whose sides are 4, 8, and 9. Hence by the Pythagorean Theorem,

$$4^2 + 8^2 = 9^2,$$
$$80 = 81.$$

Is he convinced by this that his value of π is incorrect? Not he. He concludes that *the Pythagorean Theorem is false*. This is a perfect illustration of the futility of arguing with a pi-maker. Whenever you get one backed into an absolutely inextricable corner, he simply disappears through the wall to mock you from the other side.

Neither 1935 nor any other year to date has seen the last of the pi-sters. On my desk as I write is a pamphlet dated 1950, entitled 'The True Value of π and the Fallacy of Archimedes.' The 'true value,' it develops, is 3.1547 · · ·, and weird and wonderful is the proof thereof. Along with the pamphlet (for which I was fleeced of eighty cents—but it was worth it) came an advertisement of further important literature of the same ilk, in which is voiced the complaint: 'But my program is fiercely fought by the conservative mathematicians, who feel humiliated by it; and by the publishers of science, who fear for their business interests, as if their best foundation were not in truth and honesty.' The tone of righteous indignation is typical of the misunderstood martyrs who carry the torch for π.

Before leaving this odd and most un-mathematical subject, I cannot refrain from quoting a passage from an article by N. T. Gridgeman: *

It is sad that de Morgan did not live to appreciate the wondrous things that were to happen in the Middle West. In 1889, Dr. Goodwin, a medical man of Solitude, Indiana, copyrighted the following statement: 'A circular area is equal to the square on a line equal to the quadrant of the circumference; and the area of a square is equal to the area of the circle whose circumference is equal to the perimeter of the square.' Later he published an article *in an orthodox mathematical journal* on the 'The Quadrature of the Circle,' the opening paragraph being the copyrighted statement given above and that thereafter drivelled on to 'prove' that $\pi = \frac{16}{5}$ by a method that involves, by implication, the identity $7^2 = 50$. The article was published 'at the request of the author' and is significantly never referred to in the lively discussion columns of subsequent issues. Dr. Goodwin made good use of that publication; it enabled him to claim that his quadrature was 'accepted as a contribution to scientific thought' by the journal, a claim that was prominent in the taking of his next step. On January 18, 1897, Goodwin's county representative introduced House Bill No. 246 into the Indiana State Legislature; it began:

A bill for an act introducing a new mathematical truth and offered as a contribution to education to be used only by the State of Indiana free of cost by paying any royalties whatever on the same, provided

* N. T. Gridgeman, 'Circumetrics,' in *The Scientific Monthly*, Vol. 77, No 1. (July 1953), pp. 31–5. I am indebted to the American Association for the Advancement of Science, who hold the copyrights, for permission to reproduce the passage. From the same article I have already borrowed the descriptive term 'pi-makers.'

it is accepted and adopted by the official action of the legislature of 1897.

> Section 1. Be it enacted by the General Assembly of the State of Indiana: It has been found that a circular area is to the square on a line equal to the quadrant of the circumference, as the area of an equilateral rectangle is to the square of one side. Etc., etc., etc.

The worthy Representatives, finding that they would be able to collect royalties from the willing scholastic converts beyond their borders, routed the bill through the Committee on Canals (*sic*) then through the Education Committee, both of whom approved it, and on February 2 the house voted its acceptance by 67 to 0. It then went to the Senate who thoughtfully referred it to the Committee on Temperance, which, evidently convinced that it did not endanger the sobriety of Indiana's residents, promptly passed it. By this time, however, the Indianapolis press had got hold of the story, and it began to spread across country. Just in time the Senate woke up to the enormity of the gaffe it was about to commit, and on February 11, House Bill No. 246 was postponed indefinitely. So fell the curtain on what is surely one of the most fantastic scenes in the history of both pi-istics and legislature.

• • •

Having devoted so much space to the pi-makers, we should do their colleagues, the angle-trisectors, the favor of at least a casual mention. Every schoolboy who has a liking for geometry and a spark of gumption spurns his teacher's advice and tries to trisect an angle with straightedge and compass. After some unsuccessful attempts he puts the problem aside; and eventually he acquires enough mathematics to follow the proof of the impossibility of such a trisection. But your dyed-in-the-wool angle-trisector never gets beyond the schoolboy stage, and every year some weary mathematician is asked to examine a new trisection and challenged to 'Find the flaw, if you think there is one.' Naturally it is useless to tell the trisector that it is known *a priori* that there *has* to be a flaw. The proof that an arbitrary given angle cannot be trisected in the classical sense is much easier than the proof that the circle cannot be squared, but it is still hard enough to be out of the range of the complete amateur.

The trisection is impossible if we are to use the ruler for drawing straight lines *only*. This is always the understanding in a ruler-and-compass construction. If, however, the ruler can be used as a measuring device, then a trisection is easy. Archimedes gave the solution of Figure 66. Given any angle *COD*, extend one side of it as shown. Draw a semicircle of any radius (r), center at *O*. Mark the edge of a ruler with two points *A* and *B* whose distance apart is r. Lay the ruler across the semicircle in such a position that *A* is on the extended side of the angle, *B* is on the semicircle, and the ruler also passes through *C*. Draw a line *OX*

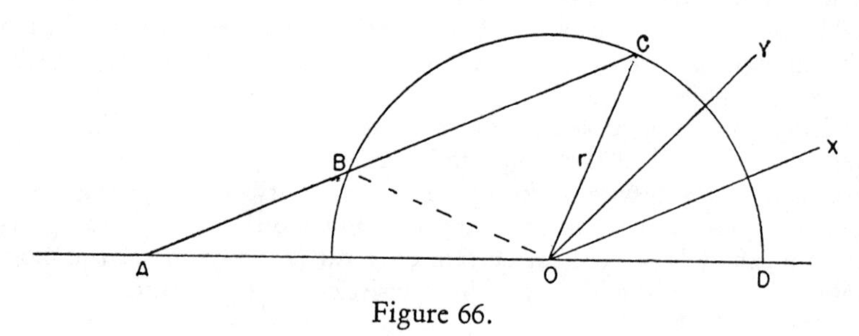

Figure 66.

through *O* parallel to *AC*. Bisect angle *COX* with *OY*. *OX* and *OY* are the required trisectors of angle *COD*. The (very easy) proof is given in the Notes.

It has recently (1955) been pointed out that for any angle θ, arbitrarily small angles ϕ exist and can be constructed with ruler and compass such that $(\theta - \phi)$ can be trisected with ruler and compass. This means that although an angle cannot be exactly trisected, it can be approximately trisected, and the approximation can be made as accurate as you please.

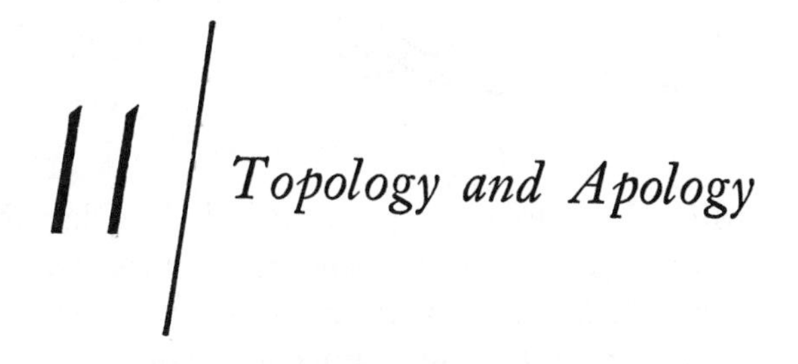

11 | *Topology and Apology*

A BRAND NEW SUBJECT has grown in the past thirty years from the status of a curio to that of a major branch of mathematics: topology. Topology is a kind of geometry, but all of its best theorems are developed by analysis. Hence most topological theorems are beyond our range here, and we shall have to be content with some rather trivial examples. One must not conclude from these elementary fragments that topologists spend their time making paper models or tying knots.

One kind of transformation that does not alter the topological character of an object is deformation. Thus a cube is topologically equivalent to a sphere, inasmuch as a solid cube made of modeling clay could be deformed into a sphere merely by 'kneading' it, without tearing or breaking its surface. It could not, however, be molded into a torus (doughnut-shaped ring) unless a hole were bored through it or two surfaces were joined together; hence it is not topologically equivalent to a torus. But a doughnut is topologically equivalent to a cream pitcher with a handle.

• • •

A topological investigation that proves to be very difficult is the theory of knots. If a simple knot (or even a complicated one) is loosely tied in one end of a long piece of rope, it is possible

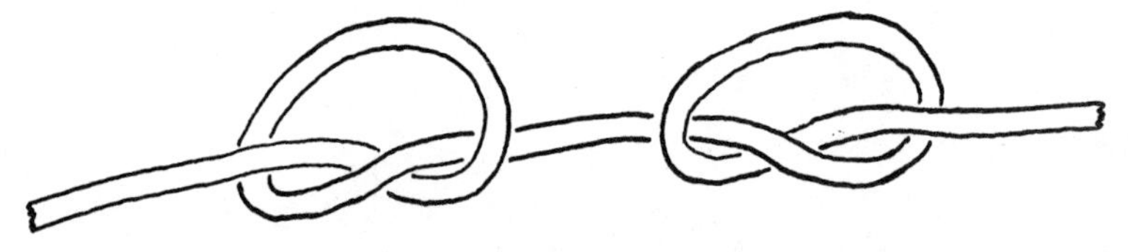

Figure 67.

to work it along the rope toward the other end. Suppose we tie a knot near each end, and then work the knots toward each other. It is assumed, from actual experiments, that it can never happen that the two knots will untie each other—cancel each other out—when they come together. But no one has succeeded in proving this statement. Figure 67 shows two knots, each the reverse of the other: one is left-handed, the other right-handed (or, if you prefer, they spiral in opposite directions). Yet they will not untie each other when brought together. Instead, one passes through the other and out the other side, leaving both unaltered.

• • •

Tie a loop of string to the end of a pencil, the loop being definitely *shorter* than the pencil (Fig. 68). This can now be attached to the buttonhole of your jacket without untying the loop. (See Fig. 69.) The operation depends on the fact that topologically the whole assembly is not fastened to the jacket at all. This is much more difficult to solve from scratch than most parlor tricks. It would be interesting to know whether this is a

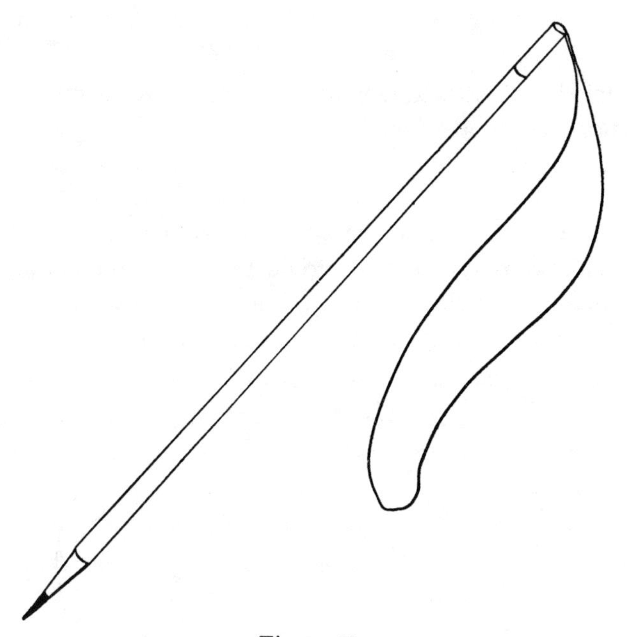

Figure 68.

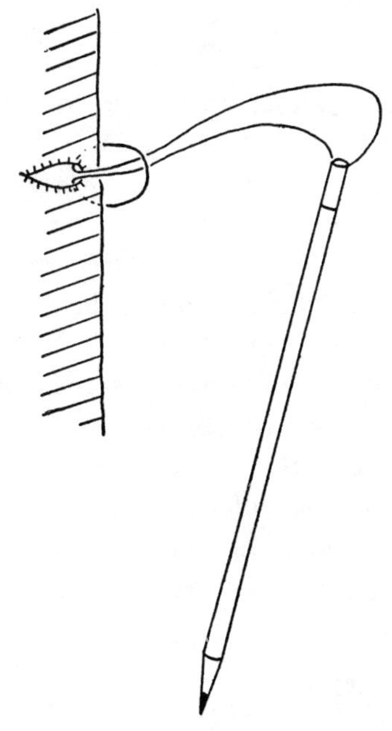

Figure 69.

hard trick because its solution is associated with mathematical notions of a fairly abstract kind.

• • •

How would you define the two sides of a surface—a piece of paper, for example? One way would be to say that a bug crawling on one side could not get to the other side without either piercing the paper or going over an edge. Another way might be

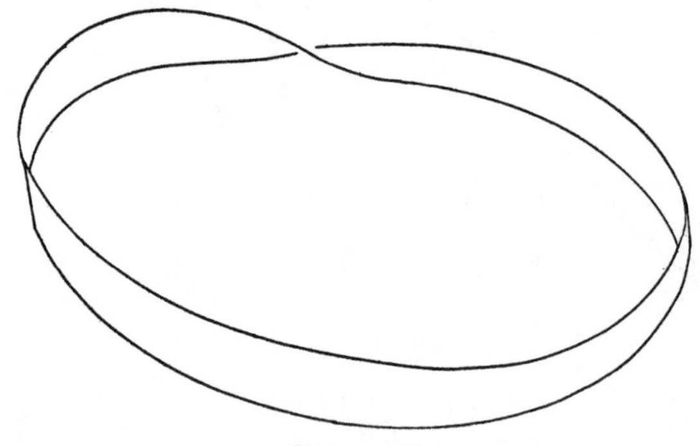

Figure 70.

to start painting and paint all the surface you could get to without going over an edge. The painted surface would then constitute one side. These two definitions are of course the same.

The geometrician A. F. Moebius discovered that under this or any other reasonable definition he could easily construct a surface with only *one* side. You can, too. Take a long narrow strip of paper, give it a half-twist, and paste the ends together. You have what is known as a Moebius strip, and a bug could reach *all parts* of its surface without going over an edge. (Fig. 70.)

Among other inventions of this kind is Klein's Bottle: it has no inside and no outside, and yet it would serve to carry wine in. Which is more intoxicating, the wine or the idea of the bottle?

• • •

Suppose we draw a map with the requirement that no two countries with a common border may be the same color. (A single point does not count as a common border.) No map representing countries on the surface of the earth, however complicated, could ever require more than four colors. In Figure 71, *A* is now completely surrounded, so that the next country can be colored with *A*'s color. No satisfactory proof of this famous four-color theorem is known, although most mathematicians surmise that it must be correct.*

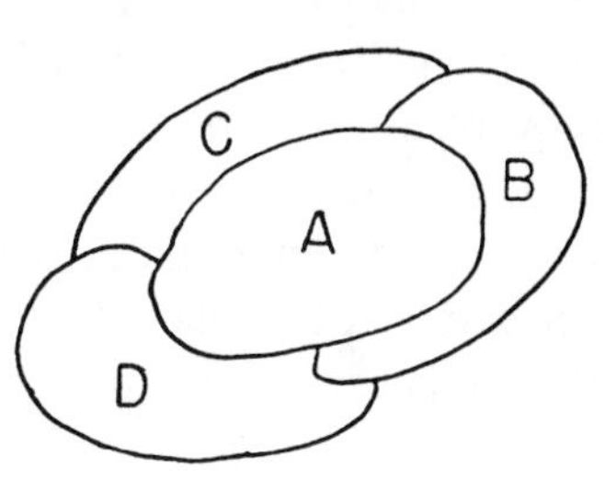

Figure 71.

If the map is drawn on the surface of a torus, more colors are needed. Seven in fact *may* be needed, but no more than seven *can* be needed. This seemingly more difficult seven-color theorem for the torus has been adequately proved.

The theorem has no analogue in three dimensions. If we ask how many different-colored stones would be needed to build a mound, if no two stones of the same color could have any part of a surface in common, and if we are allowed to use stones of any shape whatever, then the answer is that we might need an unlimited number of colors.

• • •

Topology is one of the 'going' fields in mathematical research today. The subject is being enlarged so rapidly that a specialist has difficulty keeping abreast of only the new *terms* being introduced, to say nothing of the content of the theorems. Perhaps organization and synthesis of all this new material will take place eventually. At present the subject is in the mushrooming stage characteristic of a new branch of mathematics. Development follows development so fast that one can scarcely tabulate all the results.

Does topology have any applications to so-called 'real life'? Very few. But it is much too soon to do any evaluating. It has

* A very long and involved proof by computer has been completed. [1994 note.]

always been a mistake to ask of a new mathematical creation, 'What good is it?' Topology has already made important and far-reaching contributions to other branches of mathematics; it has not had time to do much contributing to the other sciences. If you insist on 'practical' applications, we can offer one inconsequential one, and one of a more serious nature.

A theorem credited to L. E. J. Brouwer, the contemporary Dutch mathematician, states that it is impossible to parametrize the surface of a sphere (or spheroid) with any kind of co-ordinate system (network of lines) in terms of which direction has meaning at every point: there must always be at least one vortex, or whorl-point, at which direction is undefinable. (For example, the north pole is such a point in terms of the earth's co-ordinate system; from there, every direction is south.) An amusing consequence of this theorem is that there must always be at least one point on the surface of the earth where, at least momentarily, there is no wind.

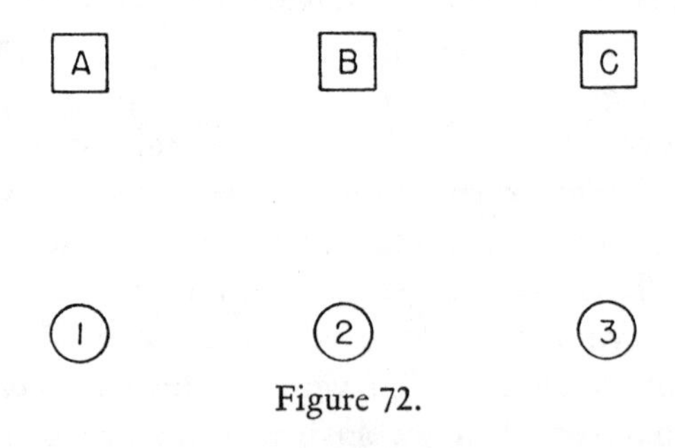

Figure 72.

The Problem of the Three Utilities depends for its solution on topological considerations. There were three eccentric neighbors (professors, no doubt) who lived in three houses *A*, *B*, and *C*. They each wanted an electric line, a water line, and a gas line direct to the respective utility plants numbered 1, 2, and 3 (Fig. 72). But each neighbor refused to allow any line to cross any other line. Could the contractors lay the lines to satisfy all three neighbors?

It should be recognized at the outset that it really makes no difference how the houses and the plants are oriented. Starting with *A*, *B*, 1, and 2, we observe that they can be connected in the required fashion (Fig. 73A) and that this connection is topologically equivalent to that shown in Figure 73B. Any other way of laying the lines will also be equivalent.

Now we need an obvious-sounding but famous theorem of Jordan: every simple closed curve in the plane has one inside and

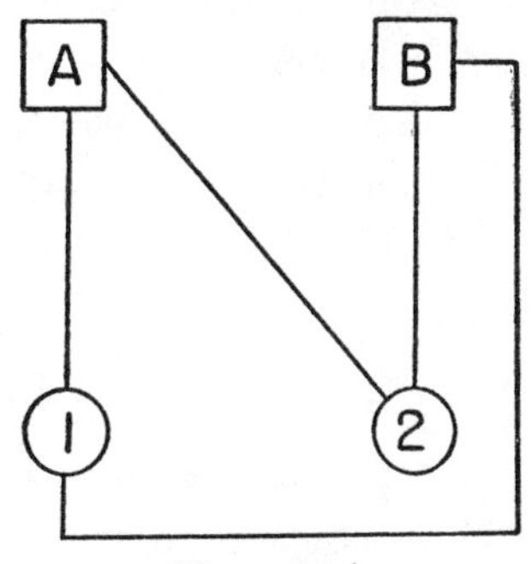

Figure 73A.

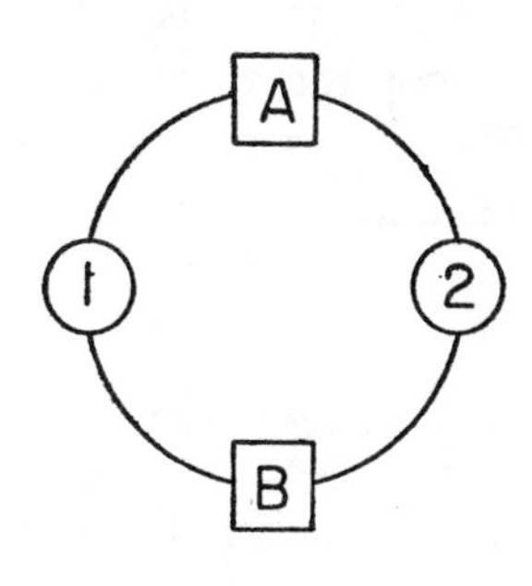

Figure 73B.

one outside. Take an ordinary elastic band and throw it down on the desk, being careful only that it forms *one* loop, and does not cross itself as in a figure 8. The band now lies in a simple closed curve, no matter what shape it takes. What Jordan's Theorem says is that you cannot pass from inside to outside without crossing the curve. Figures 73A and 73B represent two such curves if we regard the houses and plants as shrunk to points.

Suppose next that we put in the third house, *C*. If it goes inside, it can be satisfactorily connected to utilities 1 and 2. But then the inside will be *compartmented;* and no matter where you put utility No. 3, it will be cut off in terms of Jordan's Theorem from *one* of *A*, *B*, or *C*. Precisely the same thing happens if *C* is put outside.

Thus the solution is in the negative: the contractors cannot satisfy the three eccentric neighbors.

This process has serious application in the theory of stamped circuits. The circuits of complex electronic devices are so intricate that it is often simpler to stamp them out in metal than to lay insulated wires. Such stamped circuits cannot cross or they would short out. The question of what non-crossing circuits are possible is precisely an extension of the problem of the three utilities.

• • •

Until the present century all new mathematics of any stature was done by Europeans (or Orientals). It is safe to say that Americans at last have a finger in the pie. Particularly in topology important contributions are being made in this country—some by Europeans residing here, it is true, but also many by native Americans. The time is ripe; America is on its mathematical way.

• • •

We have come to the end of our tour, and I am assailed by grave doubts. Have we succeeded even partly in accomplishing what we so ambitiously set out do do? Have you seen anything at all through the mathescope? The book began at a leisurely tempo but the pace was accelerated as we progressed until, toward the end, we talked about some quite sophisticated and elusive ideas. Did you sometimes feel lost? You probably had reason to.

But may I dare to hope that you were occasionally able to get the mathescope into focus and see some of the really magnificent pictures that can appear on its screen? Perhaps you had only a brief insight, a sudden flashing view glimpsed at an odd moment, of what Mathematics with a capital M is all about. If this happened once, I am glad. If it happened more than once, it is all anyone could have asked for.

Many fine books have been written in an attempt to popularize mathematics. Some are mentioned in the Notes; and it is to be hoped that, now that you have taken tne first hurdle, you may find it your desire to read some of them. They all reflect the same thing: an overwhelming urge on the part of their authors to proclaim to the world that their beloved mathematics is not the dry-as-dust, ugly, routine affair that so very many people think it is.

We like our subject, and we would like others to like it. If I have gained any converts to the mathematical fold, may I welcome and congratulate you. Whatever further time you care to devote to the subject will be rich in intellectual rewards.

And to all of you who have stuck it out to the end, I say—thank you. I have never lectured to a more attentive audience.

LIBRARY, UNIVERSITY COLLEGE CHESTER

Notes

P. 7. The Einstein remark is from a book published some years ago in Germany, and republished (1954) under the title *Relativity: the Special and the General Theory* (Methuen & Co., London).

P. 9. For further examples of subsequent applications of 'pure' mathematics, see Jacques Hadamard, *The Psychology of Invention in the Mathematical Field* (Princeton University Press, 1949), Chapter IX.

Chapter 1. The whole of this chapter is supplemented by G. H. Hardy's *A Mathematician's Apology* (Cambridge University Press, 1940), although he sometimes overstates the case. See also Eric Temple Bell, *The Queen of the Sciences* (Williams & Wilkins Co., Baltimore, 1931), and the introduction to Morris Kline's

Mathematics in Western Culture (Oxford University Press, Inc., 1953).

P. 12. The proof of the irrationality of $\sqrt{2}$ given on pp. 12–14 has always been considered a classic example of a pure mathematical proof of the highest caliber: concise, complete, powerful. Yet Joseph Louis Lagrange (1736–1813) condensed it into a single sentence: 'It [$\sqrt{2}$] cannot be found in fractions, for if you take a fraction reduced to its lowest terms, the square of that fraction will again be a fraction reduced to its lowest terms, and consequently cannot be equal to the whole number 2' (*Lectures on Elementary Mathematics*, Open Court, p. 11). Careful reading reveals that this sentence contains all the essentials of the longer proof. It requires a mathematician of the stature of a Lagrange to dare to improve on what is generally accepted as 'perfect.'

P. 13. If p^2 is even, p is even. *Proof*. Suppose p were odd. Then it would be of the form $2k + 1$, k an integer. Therefore its square would be $p^2 = (2k + 1)^2 = 4k^2 + 4k + 1$. But now $4k^2$ and $4k$ are surely even, no matter what k is. Hence p^2 would be an odd number (even + even + 1), a contradiction. What we have proved is that if p^2 is even, p cannot be odd.

P. 15. The quotation is from de Morgan's *A Budget of Paradoxes* (Open Court, 1915), Vol. II, p. 210. Sometimes long-winded, often greatly exercised over issues that were important to the English intelligentsia of 100 years ago but seem trivial today, the *Budget* none the less rewards the reader with frequent shafts of dry wit and pungent sarcasm. One pictures de Morgan as a man who may have had his foibles, but who walked through life with dignity and honor, the very prototype of the nineteenth-century Old World gentleman and scholar.

P. 15. For this opinion on the Pythagorean reaction to $\sqrt{2}$, see Vera Sanford, *A Short History of Mathematics* (Houghton Mifflin, 1930), p. 183.

P. 18. For 'The Geometry of the Square Root of Three,' see the author's paper by this title in the *American Mathematical Monthly*, Vol. 56 (1949), p. 172.

P. 21. The sum of a geometric progression.

Let S_n be the sum of the first n terms and take $r > o$:

$$S_n = a + ar + ar^2 + \cdots + ar^{n-2} + ar^{n-1}.$$

Now multiply both sides of this equation by r:

$$S_n r = ar + ar^2 + ar^3 + \cdots + ar^{n-1} + ar^n.$$

Subtract the second equation from the first:

$$S_n - S_n r = a - ar^n.$$

Solving this for S_n,

$$S_n = \frac{a - ar^n}{1 - r}.$$

Now if $r < 1$, $1/r > 1$, or $1/r = 1 + k$, k a fixed positive quantity. By the binomial expansion,

$$(1 + k)^n = 1 + kn + \cdots + k^n.$$

Hence as $n \to \infty$, examination of the first two terms alone is sufficient to show that $(1/r)^n = (1 + k)^n \to \infty$, in other words $r^n \to 0$. (The argument of the last few lines is customarily and incorrectly omitted in the textbooks, being replaced by the statement, by no means evident, that r^n can be made arbitrarily small by taking n sufficiently large.) Thus if S means the limit of S_n, then

$$S = \frac{a}{1 - r}.$$

P. 23. In his generally excellent book, *Mathematical Recreations* (Dover, 1953), Maurice Kraitchik makes a slip in discussing cyclic numbers. For some reason he states (p. 76) that 142857 is the only such number. In a letter to the author, Mr. Kraitchik has explained that he did not mean to consider numbers beginning with a zero on page 76. (Unfortunately he says the opposite on page 75.) Even with this correction, the last sentence of page 76 is false.

The interested reader is referred to Solomon Guttman, 'On Cyclic Numbers,' *American Mathematical Monthly*, Vol. 44 (1934), p. 159. For a tabulation of the latest work done on the subject, see the reference list at the end of K. Subra Rao's note in the same periodical, Vol. 62 (1955), p. 484.

P. 24. Euclid's proof that the number of primes is infinite: Suppose there is a largest prime, P. Now $P! + 1$ is not divisible by P or by any lesser number (because $P!$ is divisible by all of them). Hence $P! + 1$ is either itself a prime, or it is divisible by some prime greater than P. Both possibilities contradict the assumption that P is the greatest prime. Hence there is no greatest prime, which is another way of saying that the number of primes is infinite.

P. 24. The prime number theorem. The approximate number of primes less than any number n is equal to n divided by the logarithm of n. The truth of this important theorem had been guessed early in the last century by Carl Friedrich Gauss, one of the greatest mathematicians of all time. Its proof was not developed, however, until near the beginning of the present century, and then only with the aid of difficult methods employing some of the most powerful tools of modern analysis. Consequently the announcement (in 1949) that an elementary—but not easy—proof of the theorem had been devised by a young mathematician at Syracuse University (N. Y.) caused a sensation in the mathematical world. (Atle Selberg, 'An Elementary Proof of the Prime Number Theorem,' *Annals of Mathematics*, Vol. 50, pp. 305–13.)

The kind of logarithm used in the prime number theorem is not the common or Briggs logarithm to the base 10, but the natural or Naperian logarithm to the base e. The number e is one of the fundamental constants of mathematics, turning up again and again, like π, in the most unexpected places. Here is one definition of e:

$$e = 1 + \frac{1}{1!} + \frac{1}{2!} + \frac{1}{3!} + \frac{1}{4!} + \frac{1}{5!} + \cdots.$$

This implies that the series *converges*—has a limit. 'One can easily see' (and it is not hard to prove) that the series of inverse factorials is term by term less than or equal to the geometric progression

$$2 + 1 + \tfrac{1}{2} + \tfrac{1}{4} + \tfrac{1}{8} + \tfrac{1}{16} + \cdots$$

which we know converges. This 'comparison test' shows that the series for e itself converges. The sum, to five decimal places, is $e = 2.71828\cdots$. Another expression for e is

$$e = \lim_{n \to \infty} \left(1 + \frac{1}{n}\right)^n,$$

which reduces, under proper treatment, to the first definition.

P. 24. The proof of the existence of arbitrarily large intervals devoid of primes is adapted from Tobias Dantzig, *Number, the Language of Science* (Macmillan, 1939), p. 267.

P. 25. The number of digits in large factorials is given in Salzer's *Tables of N!*, National Bureau of Standards, Applied Mathematics Series 16, Washington (1951).

P. 26. There are no known prime Fermat numbers greater than that given by $n = 4$. The information on the handful of known composite ones is meager. These numbers are so huge that it is useless to try to study them except with the aid of very specialized methods of number theory. The available data have been tabulated by M. Kraitchik on p. 74 of *Mathematical Recreations*.

P. 27. The lines of verse are from the *Budget*, Vol. II, p. 21.

P. 27, footnote. Fermat's last theorem has been *checked*, rather than proved, for all values of n up to 2000, with the aid of an electronic calculator. Lehmer, Lehmer, and Vandiver, 'An Application of High-Speed Computing to Fermat's Last Theorem,' *Pro-*

ceedings of the National Academy of Sciences, Vol. 40 (1954), No. 1, pp. 25–33.

P. 28. The Courant and Robbins quotation is from *What Is Mathematics?*, p. 31.

P. 29. Horace Levinson, *The Science of Chance* (Rinehart & Co., 1950), p. 44.

P. 30. The Pascal whose name is attached to the Triangle is the same Blaise Pascal (1623–62) known to the general reader for his philosophical and religious writings. He produced some absolutely top-drawer mathematics, and doubtless could have done much more had he chosen to leave philosophy and religion alone. This choice, so easy today, was almost impossible to a thinker of the seventeenth century, when science was still enshrouded in the mysteries of metaphysics.

P. 31. Every bridge player should know how to use the Pascal Triangle. If you are addicted to this game, it will be profitable for you to memorize the first 5 or 6 rows. They answer the important question, what is the probability that the outstanding cards in a suit will 'break'?

Suppose you and dummy hold 9 trumps; one line of play requires the opponents' 4 trumps to break 2-2; should you take the chance, or try a different line of play? Many (inexperienced) players think that the odds are in favor of a 2-2 break. The 4 line of the Pascal Triangle tells us that this is not so. There are 16 different layouts of the cards, of which only 6 give the 2-2 break. Since the odds are thus 10 to 6 against it, another line of play should be sought.

Whenever there is an even number of trumps outstanding, the probability of their being evenly divided is one of those listed on page 34. Thus the 3-3 break of six adverse trumps is even less likely: the chances are 44 to 20 against it. But if there are only *five* out, because there are now *two* middle numbers in the row, the chances are 20 to 12 in favor of a 3-2 break.

Note that we said nothing about dropping singleton Kings,

doubleton Queens, etc., in the above discussion. If the even division is needed for the purpose of a *drop*, and not just to collect adverse trumps without an extra lead, the situation is altered, because some of the uneven divisions will include the favorable drop.

Actually the probabilities quoted above are not quite accurate for another reason. The fact that there have to be other cards in the hands, but only 52 altogether, affects the odds. Any bridge book lists the exact odds. They are close enough to those given by the Pascal Triangle so that no essential differences in play result if the simple Pascal odds are consulted.

P. 31. Some of the interesting properties of the numbers forming Pascal's Triangle are given in the author's note on 'The Binomial Coefficients,' *American Mathematical Monthly*, Vol. 57 (1950), p. 551.

P. 31. Theorem. All the coefficients in the expansion of $(a + b)^n$ except the first and last are divisible by n if and only if n is prime.

Proof. Sufficiency. $\dfrac{n!}{r!(n - r)!} = j$, an integer. Therefore $n! = j \cdot r!(n - r)!$. Neither $r!$ nor $(n - r)!$ is divisible by n if n is prime. But $n!$ is divisible by n. Hence j must be divisible by n.

Necessity. If n is composite, let k be its smallest prime factor. The $(k + 1)$st term of the expansion is

$$j = \frac{n(n-1)\cdots(n-k+1)}{k!}.$$

None of the factors $(n - 1), \cdots, (n - k + 1)$ is divisible by k. Therefore to reduce j to an integer, k must be divided into n. The result is an integer:

$$j = \frac{(n/k)(n-1)\cdots(n-k+1)}{(k-1)!}.$$

Now n does not divide this j; for none of $(n - 1), \cdots, (n - k + 1)$ contains k to start with, and division by the factors of $(k - 1)!$ can never introduce k by composition because k is a prime.

P. 31. Births in the United States average about 51 per cent boys, according to figures in the *World Almanac*.

P. 33. Dr. Weaver was very definitely right.

P. 34. Here is a proof, which will have to be passed over by the non-mathematician, that the probability of tossing exactly half heads in $2n$ tosses of a coin tends to zero as n is increased indefinitely.

The probability of getting exactly half heads in $2n$ tosses is

$$p = \frac{n!}{(n - n/2)!(n/2)!} \div 2^n = \frac{n!}{2^n(n/2)!}.$$

By Stirling's Approximation, this reduces, for large n, to $p \simeq \sqrt{2/(\pi n)}$. Thus by making n suitably large, p can be made arbitrarily small.

P. 34. For a detailed discussion of this 'game,' classically known as the St. Petersburg Paradox, see Kraitchik, *Mathematical Recreations*, pp. 135–40.

P. 36. Birthdays. Let p be the probability that 25 birthdays are *different*.

$$p = \frac{365}{365} \cdot \frac{364}{365} \cdot \frac{363}{365} \cdots \frac{341}{365} < \frac{1}{2}.$$

Since $p < \frac{1}{2}$, the probability that they are *not* all different is greater than $\frac{1}{2}$.

In order to get an approximate value of the product of the 25 fractions, use the 'average' of the numerator factors, 353: $(\frac{353}{365})^{25} \simeq .41$. This is a little too small, but not much. The actual value is substantially under $\frac{1}{2}$, and as the number of people increases, the probability of all birthdays being different decreases rapidly. For 22 people it is just over $\frac{1}{2}$, for 23 people, just under $\frac{1}{2}$. For 40 people, it is about $\frac{1}{5}$.

For a more accurate method of determining these probabilities, see William Feller, *Probability Theory and Its Applications* (Wiley, 1950), p. 29.

P. 39. Somewhere back in your third-grade arithmetic book, probably in small print at the bottom of the page, you were warned against attempting to divide by zero. Why? We could answer simply by defining division as a process that does not include zero among possible divisors. We can do a little better than that, however. Division is usually defined as the inverse of multiplication. That is, $10 \div 2$ means some number which when multiplied by 2 will give 10. 5 does it. Following this definition, $10 \div 0$ would mean some number which when multiplied by 0 will give 10. There exists no number that does it.

Some of the older texts wrote $\frac{10}{0} = \infty$. This is a very dubious procedure, leading to various algebraic pitfalls. The symbol ∞ is not a number, and does not obey the ordinary rules of algebra. You may write ∞ if you know what you are doing; but it is far safer to say that division by zero is simply not allowed.

The proper way to write the alleged 'equation' $\frac{10}{0} = \infty$ is:

$$\lim_{x \to 0} \frac{10}{x} = \infty;$$

and the proper interpretation of *this* equation is: for any preassigned arbitrary number N, however large, there exists a value of x *greater than zero* such that $10/x$ is greater than N. It is by such precise statements as this that one starts on the road toward the conquest of the infinite. We can afford to brook no nonsense about dividing by zero.

The Agnew reference is to *Differential Equations* (McGraw-Hill, 1942), p. 35.

P. 39. The line with the square brackets says $[\frac{1}{2}]^2 = [-\frac{1}{2}]^2$. This is correct. However, taking the square root of both sides leaves $\frac{1}{2} = -\frac{1}{2}$, which is most certainly not correct. You cannot take the square root of both sides of an equation without first inspecting for the possibility of sign trouble.

This 'proof' was found written on the wall of the hallway in the math building at Cornell University. Underneath it, someone had penciled 'Extremely improbable.' This might be taken to represent the apotheosis of mathematical open-mindedness: the writer seems to be saying, 'I am intuitively certain that 1 does not equal 0; but if you can prove to me that it does, by a flawless line of logical reasoning, I shall be forced to accept your proof!'

The exercise is also given in Kasner and Newman, *Mathematics and the Imagination* (Simon & Schuster, 1940), p. 209.

P. 39. To qualify, the racers must do the two laps in two minutes, for an average of a mile a minute. But the driver who averages 30 m.p.h. for the first mile has already used up his allowance of two minutes (30 m.p.h. $= \frac{1}{2}$ mile per min.). Therefore he would have to do the second lap in nothing flat.

In general, you cannot average rates over equal *distance* intervals; they can be averaged only over equal *time* intervals. If a car goes 30 m.p.h. for one *minute*, and then 90 m.p.h. for one *minute*, it averages 60 m.p.h. for the two minutes and in fact covers two miles.

A somewhat different difficulty is encountered in the problem of the two apple women. Each had 30 apples to sell; one was going to sell 3 for 1¢, and the other 2 for 1¢. This would have netted them a total of $10 + 15 = 25$ cents. But they decided to pool their resources, and sold all 60 at 5 for 2¢, which looked like the same price. But this netted them only 24¢. The moral is obvious: those who engage in private enterprise should beware of the perils of syndicates.

P. 40. Not every iterated radical works out so neatly. The expression $\sqrt{y + \sqrt{y + \sqrt{y + \cdots}}}$ has a limit which is an *integer* x if and only if $y = x(x - 1)$. Thus $30 = 6 \times 5$, and therefore $6 = \sqrt{30 + \sqrt{30 + \sqrt{30 + \cdots}}}$. For proof see problem E-874, *American Mathematical Monthly*, Vol. 57 (1950), p. 186.

P. 41. The distributive law: $a(b + c) = ab + ac$.

P. 41. The argument that the product of two negative quantities is a positive quantity is adapted from Birkhoff and MacLane, *A Survey of Modern Algebra* (Macmillan, 1953), p. 5.

It is irrelevant (but not coincidental) that in grammatical usage two negatives make an affirmative. One only courts trouble in trying to explain anything mathematical by grammatical analogues (even though Mr. Lancelot Hogben thinks otherwise). For example, a careless thinker may identify 'zero' with 'nothing.' This identification is usually false; zero is a perfectly respectable number. A line that has zero slope is horizontal; a line that has no slope is vertical (its slope is not defined).

P. 42. $$\frac{1}{a^2} = \frac{a^n}{a^n} \cdot \frac{1}{a^2} = \frac{a^n}{a^{n-2}} = a^{n-n-2} = a^{-2}.$$

P. 46. The differential equation of the generalized wine-and-water problem is

$$\frac{dw}{dt} = \frac{100 - w}{100} - \frac{w}{100}$$

(rate of change = rate of gain − rate of loss).

P. 50. L. E. Dickson, in his monumental *History of the Theory of Numbers* (Chelsea, 1952), Vol. II, pp. 582–5, states that there are many solutions to the problem of finding a cube that is the sum of three cubes of numbers in arithmetic progression; but that $3^3 + 4^3 + 5^3 = 6^3$ is the only one in which the three cubes are consecutive.

P. 50. The references are: W. F. White, *A Scrap-Book of Elementary Mathematics* (Open Court, 1910), p. 61; and Yancey and Calderhead, 'New and Old Proofs of the Pythagorean Theorem,' *American Mathematical Monthly*, Vols. 3–6 (1896–9).

P. 51. All primitive Pythagorean number triples a, b, c are given by the following formulas, where u and v run through all the integers, with the restrictions that $v > u$, and that u and v have no common factor, and that u and v are not both odd:

$$a = v^2 - u^2; \qquad b = 2uv; \qquad c = v^2 + u^2.$$

For discussion of this theorem see, for instance, Courant and Robbins, *What Is Mathematics?*, p. 41. (Note, however, the misprint at the bottom of the page: all *u*'s and *v*'s in the last three lines should be interchanged.)

P. 52. Figure 8 follows H. E. Dudeney, *Amusements in Mathematics* (Thos. Nelson & Sons, Ltd., London, 1917), p. 32.

Figure 9 is given by Edouard Lucas in *Récréations Mathématiques* (1883), Vol. II, p. 130, with the following note: 'Le Numéro de janvier 1882 de *The mathematical Magazine* contient une variente de cette démonstration, attribuée à Garfield, l'infortuné président des Etats-Unis, assassiné l'année précédente.' Garfield's proof (loc. cit. Vol. I, No. 1, p. 7) is less simple than the proof of our Figure 9.

For another spectacular proof, see page 2 of Hugo Steinhaus, *Mathematical Snapshots* (Oxford University Press, Inc., 1950).

P. 52. What Descartes invented is now called Analytic Geometry. In the estimation of E. T. Bell, whose *Men of Mathematics* (Simon & Schuster, 1937) is a classic in the field, 'Descartes did not revise geometry; he created it' (p. 54).

P. 53. This example is from Felix Klein, *Elementary Mathematics from an Advanced Standpoint*, page 47 of Vol. I of the English translation (Dover, 1945). There is a description of a class of circles having the same property in the *American Mathematical Monthly*, Vol. 56 (1949), p. 407.

P. 54. For further discussion of the 'hole through the cube,' see the *American Mathematical Monthly*, Vol. 57 (1950), p. 339.

P. 57. This proof is somewhat abbreviated. For a fuller treatment, see Hilbert and Cohn-Vossen, *Geometry and the Imagination* (Chelsea, 1952), p. 89.

P. 59. It is obvious from Figure 74, which shows one of the pieces, what the other one looks like and how they slide together. The 'important mathematical principle' is never to assume more than what is given. Do not let your mind fill in what your eye does not see. This puzzle is from Dudeney, *Amusements in Mathematics*, p. 145.

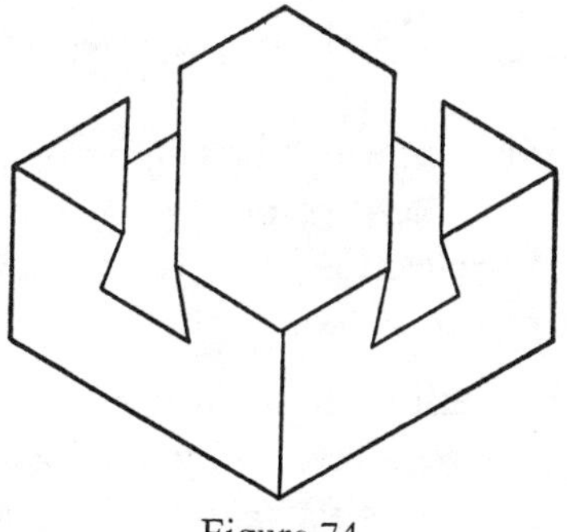

Figure 74.

P. 64. Proof that light emanating from the focus of a paraboloid is reflected in a beam parallel to the axis of the paraboloid.

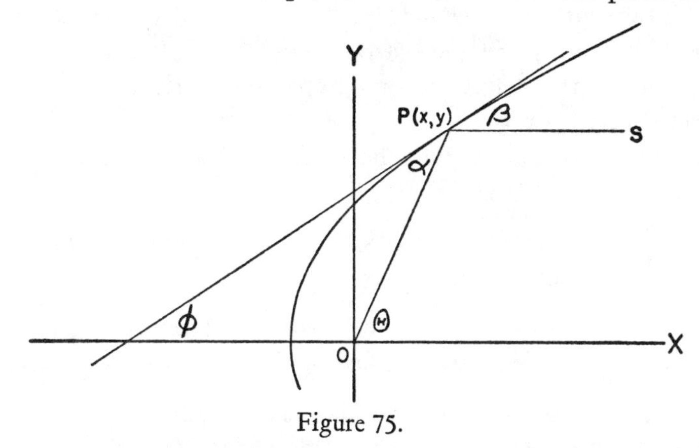

Figure 75.

Take the equation of a plane section through the axis of revolution in the form which places the focus at the origin:

$$y^2 = 2px + x^2.$$

Differentiating: $\tan \phi = dy/dx = p/y.$

Hence: $\tan 2\phi = \dfrac{2 \tan \phi}{1 - \tan^2 \phi} = \dfrac{2py}{y^2 - p^2}.$

Now solving the equation of the parabola for x gives $x = \dfrac{y^2 - p^2}{2p}$.
Hence $\tan \theta = y/x = \dfrac{2py}{y^2 - p^2} = \tan 2\phi$. Therefore $\theta = 2\phi$. But $\theta = \phi + \alpha$ (exterior angle = sum of two opposite interior angles). Therefore $2\phi = \phi + \alpha$, or $\alpha = \phi$. But $\alpha = \beta$ (angle of incidence = angle of reflection). Therefore $\beta = \phi$, which says that PS is parallel to the X-axis.

P. 70. Proof that an ellipse looks like a circle if and only if it is viewed from a point on a certain hyperbola.

The sphere of Figure 29 is tangent to the plane of the ellipse at a focus (say F_1 of Fig. 30). Let V_1 be that vertex of the ellipse on the same side of its center as F_1, and let V_2 be the other vertex of the ellipse. If P is the vertex of the cone, let PV_1 and PV_2 be tangent to the sphere at points Q_1 and Q_2 respectively. Then $PV_2 - PV_1 = Q_2V - Q_1V = F_1V_2 - F_1V_1$, *constant* for all P. This is the definition of a hyperbola with foci at V_1 and V_2 and a vertex at F_1.

This proof was first published in the *American Mathematical Monthly*, Vol. 59 (1952), p. 557.

P. 71. What happens to the surface of the liquid if the tilted container is a hemispherical cup?

P. 71. The quotation is from 'Dates of Stonehenge,' by V. Gordon Childe, *The Scientific Monthly*, Vol. 80 (1955), p. 283.

P. 72. For a discussion of the suspension bridge from the engineering standpoint, see H. M. Dadourian, *Analytic Mechanics* (Van Nostrand), p. 107. It is pointed out that if the vertical rods carrying the roadbed are constructed of lengths such that the supporting cable *must* hang in a parabola, then the load on each rod is the same if the rods are equally spaced.

For the free surface of a rotating liquid see Agnew, *Differential Equations*, p. 78.

The gas-law curves take the form $xy = k$, equilateral hyperbolas whose asymptotes are the co-ordinate axes.

Most books on analytic mechanics include the derivation of the planetary orbits starting from Newton's law of gravitation. See for instance Herbert Goldstein, *Classical Mechanics* (Addison-Wesley, 1951), p. 76.

P. 73. Proof that the creases envelop a parabola (see Fig. 76).

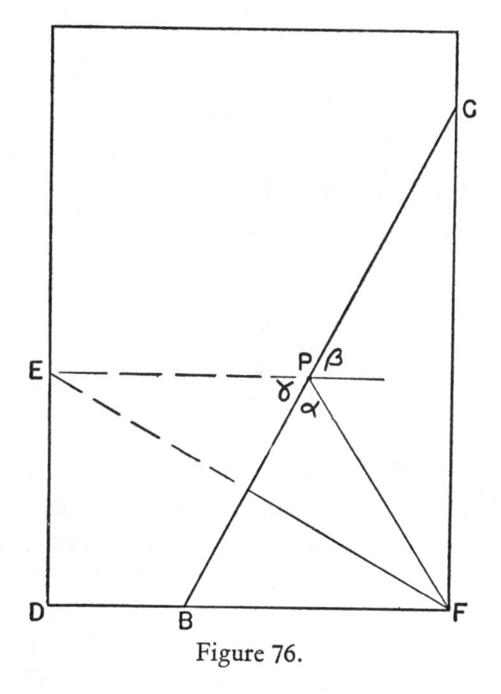

Figure 76.

Folding along BC implies symmetry with respect to BC. Hence for any position of E, if EP is parallel to DF, the point P traces out a parabola, focus at F, directrix DE (because $EP = FP$). Furthermore, $\alpha = \gamma = \beta$. Hence BC is tangent to the parabola at P, by the reflection property.

P. 75. The derivation of the parametric equations of the cycloid and the rectification and quadrature of an arch can be found in every textbook of elementary calculus.

P. 75. The mutual involute-evolute property of the cycloid is proved as follows.

In Figure 77, draw circle $OM'A'$ congruent to the generating circle OMA and tangent to it at its point of contact with CD. Extend MO until it cuts the new circle at M'. Then we observe that $\widehat{A'M'} = \widehat{AM} = \pi - \widehat{MO} = CD - \widehat{MO} = CD - CO = OD = A'E$. That is, $\widehat{A'M'} = A'E$. But this is true for *any* position of M; hence M' is tracing out a congruent cycloid, cusp at E, maximum at C.

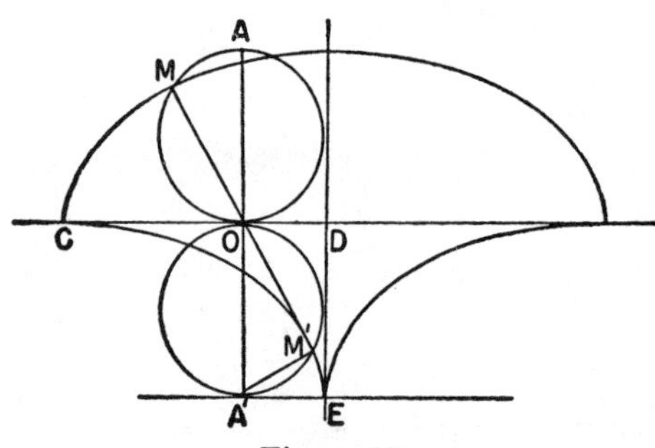

Figure 77.

Now point A is the instantaneous pivot-point of the rolling circle $OM'A'$. Hence $A'M'$ is perpendicular to the direction of motion of point M'—that is, to the tangent to the second cycloid at M'. But $A'M'$ is perpendicular to $M'O$ (any angle inscribed in a semicircle is a right angle). Hence $M'M$ is tangent to the second cycloid at M'. And $M'M$ is perpendicular to the direction of the first cycloid at M, by the same pivot property. This is what we set out to prove.

All this and more is contained in a thoroughly pleasant and pleasantly thorough treatment of the cycloid on pp. 438–40, Vol. I, of E. R. Hedrick's translation of Edouard Goursat's *Course d'Analyse* (Ginn & Co.). The three-volume 'Goursat-Hedrick,' as it is affectionately known to all mathematicians, is one of the bibles of analysis.

P. 76. The proof of the isochronous pendulum, or tautochrone property of the cycloid, is given in J. B. Reynolds' *Analytic Mechanics* (Prentice-Hall, 1929), pp. 146–8.

P. 76. The brachistochrone problem is essentially different from the tautochrone, requiring different methods of attack. For a solution see O. Bolza, *Lectures on the Calculus of Variations* (Stechert, 1946), pp. 126–8.

P. 77. Referring to Figure 36, we wish to show that P lies on ST. We know that $\angle PQC = 2\angle SOC$, because $\widehat{SC} = \widehat{PC}$ and one circle has twice the radius of the other. However, triangle PQO is isosceles, and consequently $\angle PQC = 2\angle POC$. Comparing the two equalities, we observe that $\angle SOC = \angle POC$, which says that P lies on ST.

P. 79. Felix Klein's 'Vorträge über ausgewählte Fragen der Elementargeometrie' was translated by W. W. Beman and D. E. Smith: *Famous Problems of Elementary Geometry*. It was reprinted under that title in 1955 by Chelsea Publishing Co., New York. Chapter III gives Gauss' proof. In Chapter IV, the actual construction of the 17-gon is given in 18 pages. A one-paragraph geometrical construction was developed by Richmond, whose proof (also quite short) still depends on the Gauss-Klein theory. H. W. Richmond (Cambridge, England), 'To construct a regular polygon of seventeen sides,' *Mathematische Annalen*, Vol. 67 (1909), pp. 459–61.

P. 80. See 'Area in which a narrow rod can be reversed in direction,' R. E. Greenwood, in *Pi Mu Epsilon Journal*, Vol. 1, No. 7 (1952). The Besicovitch paper is in *Mathematischen Zeitschrift*, Vol. 27 (1928), pp. 312–20.

P. 82. The 'roller property' of curves of constant width was called to my attention by Professor Walter R. Baum, of Syracuse University.

The man who knows the most about curves of constant width in this country today is undoubtedly Michael Goldberg, of the U. S. Naval Bureau of Ordnance, who has made a hobby of the subject. His paper, 'Circular-arc rotors in regular polygons,' *American Mathematical Monthly*, Vol. 55 (1948), p. 393, lists the pertinent references.

P. 83. The Appolonius pursuit problem is discussed at some length in Steinhaus' *Mathematical Snapshots*, pp. 101–6. See also *American Mathematical Monthly*, Vol. 59 (1952), p. 408.

P. 84. The patrol boat should maintain her course until she reaches the point where the trawler would be if the trawler had headed directly toward the patrol boat when the fog set in. If the trawler is not there, the patrol boat should now start out along a spiral whose origin is the point where the trawler disappeared in the fog. This spiral must (and can) be figured in such a way that, while circling the origin, the patrol boat's distance from it increases at the same rate as the trawler's. Thus the two courses must surely intersect before the patrol boat has completed one 360° circuit. In order to make the problem a reasonably practical one, the patrol boat should be capable of maintaining a speed four or five times as fast as that of the trawler. (For the technician: if the patrol boat's speed is to remain constant, the spiral must be logarithmic.)

The problem is found in Lester R. Ford, *Differential Equations* (McGraw-Hill, 1955), p. 32; and in Agnew, p. 303 (14.991).

P. 86. The loxodrome problem is stated in D. J. Struik, *Differential Geometry* (Addison-Wesley, 1950), p. 65, Ex. 11.

P. 90. For Proposition 1, Book I, and a full discussion, see T. L. Heath's translation of Euclid's *Elements*, copiously annotated (Cambridge University Press, 1908); Vol. I, p. 235, pp. 241 ff.

P. 91. For a succinct statement of the case for diagrams, see Hadamard's remarks on Hilbert's *Principles of Geometry*, in *The Psychology of Invention in the Mathematical Field*, pp. 77–8.

P. 97. *Column A*. (1) $\theta = 60°$ if the width of each side equals the width of the bottom. If the ratio of these widths is $k \neq 1$, θ for maximum cross-section becomes a somewhat complicated function of k. (2) An expression is set up for the distance from the fixed point not on the curve to any point (x,y) on the curve, and this expression is minimized. (3) The package twice as long as it is wide. (4) Height equals diameter. Paint and utility cans usually have about these proportions. Cookie cans and coffee cans often do not. Could it be that food manufacturers sometimes try to make a quart or a pound look bigger than it is by packaging their products in other than the most compact possible containers?

Column B. (1) Proved in *What Is Mathematics?*, pp. 373–5. (2) (a) A straight line. (b) A great circle. (c) A geodesic, not always easy to determine. (3) Cylindrical, circumference equaling twice the height. (4) Answered later in this chapter.

P. 99. L. L. Whyte, 'Unique arrangement of points on a sphere,' *American Mathematical Monthly*, Vol. 59 (1952), pp. 606–11. A list of unsolved problems connected with this topic is given on p. 610, loc. cit.

P. 100. For derivation of the catenary curve, see any book on differential equations; e.g. Betz, Burcham, and Ewing, *Differential Equations with Applications* (Harper & Bros., 1954), pp. 181 ff. The name comes from the Latin *catena*, a chain.

P. 102. For further details on the see-saw problem see *American Mathematical Monthly*, Vol. 60 (1953), pp. 264–5.

P. 104. H. F. MacNeish, 'Concerning the discontinuous solution in the problem of minimum surface of revolution,' *Annals of Mathematics*, Vol. 7 (1906), pp. 72 ff.

P. 105. We mention one more property of the catenary. In its simplest form, $y = \frac{1}{2}(e^x + e^{-x})$, it is the curve for which the length of a segment in linear units always equals the area under that segment in square units. More precisely, the number of units in the length of the curve from $x = a$ to $x = b$ (see Fig. 78) equals the number of units in the area bounded by the ordinates $x = a$ and $x = b$, the curve, and the X-axis. This holds for any values of a and b. Moreover, it is the only curve for which it holds, with one exception: the straight line $y = 1$. But it is clear that the location of the Y-axis is immaterial to this property. Hence there is a *family* of catenaries each of which fulfils the requirement (Fig. 79). The line $y = 1$ can be interpreted as the envelope of this family. Indeed this is no coincidence. The prop-

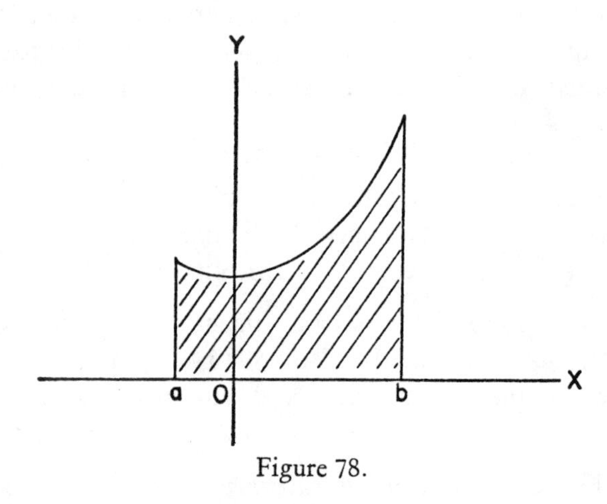

Figure 78.

erty is expressible as a first-order differential equation whose solution is the family of catenaries. Any property so expressible usually leads to a family and its envelope.

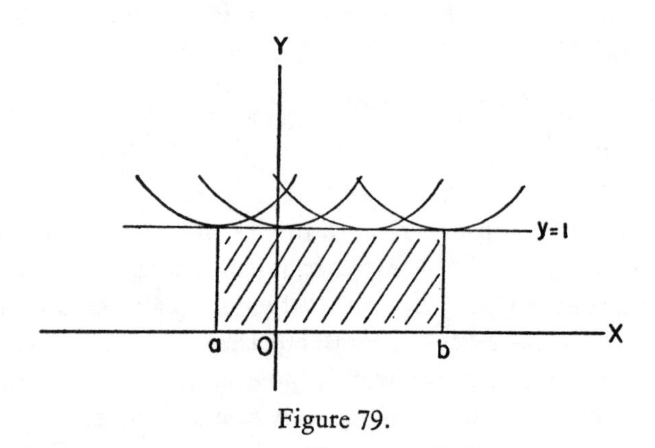

Figure 79.

An amateur faced with this problem would soon observe that a rectangle one unit high has the same area as the length of its base (top). It is a safe assumption, however, that he would probably not hit upon the general (catenary) solution.

P. 105. The soap-bubble problem is from *Mathematical Snapshots*, p. 183.

What would happen if the pipe had more than two bowls?

P. 107. The 'poem' is the work of Gerald R. Daly of the State Department, currently stationed in Hong Kong, at a safe distance from his mathematical friend.

Chapter 10. A summary of available information on the quadrature of the circle, with copious references, is found in W. W. Rouse Ball's *Mathematical Recreations and Essays* (Macmillan), pp. 293–306.

P. 108. Either by developing the MacLaurin Series or by simple long division,

$$\frac{1}{1+x^2} = 1 - x^2 + x^4 - x^6 + \cdots.$$

The series represents the function over its range of convergence $-1 < x < 1$. Being a power series without negative exponents it is uniformly convergent in any closed sub-interval, and hence can be integrated termwise:

$$\arctan x = x - \frac{x^3}{3} + \frac{x^5}{5} - \frac{x^7}{7} + \cdots.$$

This one is convergent at the top end of the range, namely for $x = 1$ (alternating series, terms monotonic-decreasing to zero), and hence represents the function there:

$$\arctan 1 = \frac{\pi}{4} = 1 - \frac{1}{3} + \frac{1}{5} - \frac{1}{7} + \cdots.$$

Unfortunately the non-mathematician cannot follow this proof, while to the mathematician it is so familiar that there is not much point in showing it to him here.

P. 109. For derivation of Wallis's product for π, see *What Is Mathematics?*, pp. 509–10, or Philip Franklin, *Methods of Advanced Calculus* (McGraw-Hill, 1944), pp. 261–2.

P. 109. The continued fraction for π is derived in George Chrystal's *Algebra* (A. & C. Black, London), Vol. II, p. 516, example 1.

P. 110. A simple and ingenious proof of the Buffon theorem, requiring only elementary calculus, is given by M. E. Munroe on p. 53 of his *Theory of Probability* (McGraw-Hill, 1951).

P. 113. Here is a quick run-down of the area under the distribution curve:

$$\begin{aligned}
\text{Area} &= 2\int_0^\infty e^{-x^2}\,dx \\
&= 2\sqrt{\int_0^\infty e^{-x^2}\,dx\int_0^\infty e^{-y^2}\,dy} \\
&= 2\sqrt{\int_0^\infty\int_0^\infty e^{-(x^2+y^2)}\,dx\,dy} \\
&= 2\sqrt{\int_0^{\pi/2} d\theta\int_0^\infty e^{-\rho^2}\rho\,d\rho} \\
&= 2\sqrt{\frac{\pi}{2}\cdot -\frac{e^{-\rho^2}}{2}\Big]_0^\infty} \\
&= 2\sqrt{\frac{\pi}{4}} \\
&= \sqrt{\pi}.
\end{aligned}$$

In 1950 C. B. Nicholas and R. C. Yates published a simpler proof not requiring the transformation to polar co-ordinates. This is one more (not very earth-shaking) example of the discovery of an easier way of handling a problem which for years had been done in a standard fashion. *American Mathematical Monthly*, Vol. 57 (1950), p. 412. Note for the mathematician sufficiently interested to pursue this reference: it can be done just as well by the method of circular discs.

P. 114. The theory of complex functions is not mastered in a day. For a readable treatment of the elements of the subject, see David Raymond Curtiss, *Analytic Functions of a Complex Variable* (the second Carus Monograph, Open Court, 1943). The equation $e^{\pi i} = -1$ is obtained by letting $y = \pi$ in equation (4), p. 53, loc. cit.

P. 115. A rational value for π which is correct to *thirty* decimal places is given in E. W. Hobson, '*Squaring the Circle*,' *a History of the Problem* (a monograph republished in 1953 by Chelsea), p. 40.

P. 115. George W. Reitwiesner, 'An ENIAC determination of π and e to more than 2000 decimal places,' *Mathematical Tables and Aids to Computation*, Vol. IV, No. 29 (1950), pp. 11–15.

P. 116. Part II of Klein's *Famous Problems of Elementary Geometry* is devoted to an exposition of Lindemann's proof of the transcendence of π. Even Klein's genius for presenting difficult subjects in an elementary way, for which he was famous, cannot make the proof easy.

P. 119. A proof of the impossibility of trisecting an angle which does not even require a knowledge of trigonometry is given by Dantzig in *Number, the Language of Science*, pp. 289–91. It is handled by trigonometric methods in *What Is Mathematics?*, p. 137.

P. 120. Referring to Figure 66, $AB = BO$. Therefore $\angle DOX = \angle OAB = \angle AOB = \frac{1}{2}\angle CBO = \frac{1}{2}\angle BCO = \frac{1}{2}\angle XOC = \angle XOY = \angle YOC$.

P. 120. Arbitrarily close trisection: M. S. Klamkin, *American Mathematical Monthly*, Vol. 62 (1955), p. 584.

There are many trisection methods, all depending, of course, on devices other than ruler and compass. For the use of the limaçon to obtain the construction given above, see T. Sundara Row, *Geometric Exercises in Paper Folding* (Open Court, 1941), pp. 143–4. Gaylord M. Merriman, in *To Discover Mathematics* (Wiley,

1941), gives the construction with the conchoid of Nicomedes (pp. 129–30), and also one of the hyperbola constructions (note 7, p. 382).

P. 122. The pencil-on-the-loop trick: Grab a handful of the jacket in the vicinity of the buttonhole and push it, including the buttonhole, through the loop. If you now put the pencil through the buttonhole you will find that you are 'hooked.' Reverse the process to get unhooked.

If you ask a friend to turn his head away while you attach the pencil to his jacket in this fashion he will probably have to wear it for the rest of the evening. I have seen only one victim escape, and he was a mathematician.

P. 124. Klein's bottle: see Hilbert and Cohn-Vossen, *Geometry and the Imagination*, pp. 308 ff.

P. 125. An unexpected new application is a topological proof that the number of primes is infinite, a theorem of our Chapter 2. Harry Furstenburg, 'On the infinitude of primes,' *American Mathematical Monthly*, Vol. 62 (1955), p. 353.

P. 127. It is amusing (and significant) that the problem has an affirmative solution on the surface of a torus.

Index

A CATALOG OF SELECTED

DOVER BOOKS

IN ALL FIELDS OF INTEREST

A CATALOG OF SELECTED DOVER BOOKS IN ALL FIELDS OF INTEREST

CONCERNING THE SPIRITUAL IN ART, Wassily Kandinsky. Pioneering work by father of abstract art. Thoughts on color theory, nature of art. Analysis of earlier masters. 12 illustrations. 80pp. of text. 5⅜ × 8½. 23411-8 Pa. $3.95

ANIMALS: 1,419 Copyright-Free Illustrations of Mammals, Birds, Fish, Insects, etc., Jim Harter (ed.). Clear wood engravings present, in extremely lifelike poses, over 1,000 species of animals. One of the most extensive pictorial sourcebooks of its kind. Captions. Index. 284pp. 9 × 12. 23766-4 Pa. $10.95

CELTIC ART: The Methods of Construction, George Bain. Simple geometric techniques for making Celtic interlacements, spirals, Kells-type initials, animals, humans, etc. Over 500 illustrations. 160pp. 9 × 12. (USO) 22923-8 Pa. $8.95

AN ATLAS OF ANATOMY FOR ARTISTS, Fritz Schider. Most thorough reference work on art anatomy in the world. Hundreds of illustrations, including selections from works by Vesalius, Leonardo, Goya, Ingres, Michelangelo, others. 593 illustrations. 192pp. 7⅛ × 10¼. 20241-0 Pa. $8.95

CELTIC HAND STROKE-BY-STROKE (Irish Half-Uncial from "The Book of Kells"): An Arthur Baker Calligraphy Manual, Arthur Baker. Complete guide to creating each letter of the alphabet in distinctive Celtic manner. Covers hand position, strokes, pens, inks, paper, more. Illustrated. 48pp. 8¼ × 11. 24336-2 Pa. $3.95

EASY ORIGAMI, John Montroll. Charming collection of 32 projects (hat, cup, pelican, piano, swan, many more) specially designed for the novice origami hobbyist. Clearly illustrated easy-to-follow instructions insure that even beginning papercrafters will achieve successful results. 48pp. 8¼ × 11. 27298-2 Pa. $2.95

THE COMPLETE BOOK OF BIRDHOUSE CONSTRUCTION FOR WOODWORKERS, Scott D. Campbell. Detailed instructions, illustrations, tables. Also data on bird habitat and instinct patterns. Bibliography. 3 tables. 63 illustrations in 15 figures. 48pp. 5¼ × 8½. 24407-5 Pa. $1.95

BLOOMINGDALE'S ILLUSTRATED 1886 CATALOG: Fashions, Dry Goods and Housewares, Bloomingdale Brothers. Famed merchants' extremely rare catalog depicting about 1,700 products: clothing, housewares, firearms, dry goods, jewelry, more. Invaluable for dating, identifying vintage items. Also, copyright-free graphics for artists, designers. Co-published with Henry Ford Museum & Greenfield Village. 160pp. 8¼ × 11. 25780-0 Pa. $8.95

HISTORIC COSTUME IN PICTURES, Braun & Schneider. Over 1,450 costumed figures in clearly detailed engravings—from dawn of civilization to end of 19th century. Captions. Many folk costumes. 256pp. 8⅜ × 11¾. 23150-X Pa. $10.95

STICKLEY CRAFTSMAN FURNITURE CATALOGS, Gustav Stickley and L. & J. G. Stickley. Beautiful, functional furniture in two authentic catalogs from 1910. 594 illustrations, including 277 photos, show settles, rockers, armchairs, reclining chairs, bookcases, desks, tables. 183pp. 6½ × 9¼. 23838-5 Pa. $8.95

AMERICAN LOCOMOTIVES IN HISTORIC PHOTOGRAPHS: 1858 to 1949, Ron Ziel (ed.). A rare collection of 126 meticulously detailed official photographs, called "builder portraits," of American locomotives that majestically chronicle the rise of steam locomotive power in America. Introduction. Detailed captions. xi + 129pp. 9 × 12. 27393-8 Pa. $12.95

AMERICA'S LIGHTHOUSES: An Illustrated History, Francis Ross Holland, Jr. Delightfully written, profusely illustrated fact-filled survey of over 200 American lighthouses since 1716. History, anecdotes, technological advances, more. 240pp. 8 × 10¾. 25576-X Pa. $10.95

TOWARDS A NEW ARCHITECTURE, Le Corbusier. Pioneering manifesto by founder of "International School." Technical and aesthetic theories, views of industry, economics, relation of form to function, "mass-production split" and much more. Profusely illustrated. 320pp. 6⅛ × 9¼. (USO) 25023-7 Pa. $8.95

HOW THE OTHER HALF LIVES, Jacob Riis. Famous journalistic record, exposing poverty and degradation of New York slums around 1900, by major social reformer. 100 striking and influential photographs. 233pp. 10 × 7⅞. 22012-5 Pa $10.95

FRUIT KEY AND TWIG KEY TO TREES AND SHRUBS, William M. Harlow. One of the handiest and most widely used identification aids. Fruit key covers 120 deciduous and evergreen species; twig key 160 deciduous species. Easily used. Over 300 photographs. 126pp. 5⅜ × 8½. 20511-8 Pa. $2.95

COMMON BIRD SONGS, Dr. Donald J. Borror. Songs of 60 most common U.S. birds: robins, sparrows, cardinals, bluejays, finches, more—arranged in order of increasing complexity. Up to 9 variations of songs of each species. Cassette and manual 99911-4 $8.95

ORCHIDS AS HOUSE PLANTS, Rebecca Tyson Northen. Grow cattleyas and many other kinds of orchids—in a window, in a case, or under artificial light. 63 illustrations. 148pp. 5⅜ × 8½. 23261-1 Pa. $3.95

MONSTER MAZES, Dave Phillips. Masterful mazes at four levels of difficulty. Avoid deadly perils and evil creatures to find magical treasures. Solutions for all 32 exciting illustrated puzzles. 48pp. 8¼ × 11. 26005-4 Pa. $2.95

MOZART'S DON GIOVANNI (DOVER OPERA LIBRETTO SERIES), Wolfgang Amadeus Mozart. Introduced and translated by Ellen H. Bleiler. Standard Italian libretto, with complete English translation. Convenient and thoroughly portable—an ideal companion for reading along with a recording or the performance itself. Introduction. List of characters. Plot summary. 121pp. 5¼ × 8½. 24944-1 Pa. $2.95

TECHNICAL MANUAL AND DICTIONARY OF CLASSICAL BALLET, Gail Grant. Defines, explains, comments on steps, movements, poses and concepts. 15-page pictorial section. Basic book for student, viewer. 127pp. 5⅜ × 8½. 21843-0 Pa. $3.95

BRASS INSTRUMENTS: Their History and Development, Anthony Baines. Authoritative, updated survey of the evolution of trumpets, trombones, bugles, cornets, French horns, tubas and other brass wind instruments. Over 140 illustrations and 48 music examples. Corrected and updated by author. New preface. Bibliography. 320pp. 5⅜ × 8½. 27574-4 Pa. $9.95

HOLLYWOOD GLAMOR PORTRAITS, John Kobal (ed.). 145 photos from 1926-49. Harlow, Gable, Bogart, Bacall; 94 stars in all. Full background on photographers, technical aspects. 160pp. 8⅜ × 11¼. 23352-9 Pa. $9.95

MAX AND MORITZ, Wilhelm Busch. Great humor classic in both German and English. Also 10 other works: "Cat and Mouse," "Plisch and Plumm," etc. 216pp. 5⅜ × 8½. 20181-3 Pa. $5.95

THE RAVEN AND OTHER FAVORITE POEMS, Edgar Allan Poe. Over 40 of the author's most memorable poems: "The Bells," "Ulalume," "Israfel," "To Helen," "The Conqueror Worm," "Eldorado," "Annabel Lee," many more. Alphabetic lists of titles and first lines. 64pp. 5³⁄₁₆ × 8¼. 26685-0 Pa. $1.00

SEVEN SCIENCE FICTION NOVELS, H. G. Wells. The standard collection of the great novels. Complete, unabridged. First Men in the Moon, Island of Dr. Moreau, War of the Worlds, Food of the Gods, Invisible Man, Time Machine, In the Days of the Comet. Total of 1,015pp. 5⅜ × 8½. (USO) 20264-X Clothbd. $29.95

AMULETS AND SUPERSTITIONS, E. A. Wallis Budge. Comprehensive discourse on origin, powers of amulets in many ancient cultures: Arab, Persian, Babylonian, Assyrian, Egyptian, Gnostic, Hebrew, Phoenician, Syriac, etc. Covers cross, swastika, crucifix, seals, rings, stones, etc. 584pp. 5⅜ × 8½. 23573-4 Pa. $10.95

RUSSIAN STORIES/PYCCKNE PACCKA3bl: A Dual-Language Book, edited by Gleb Struve. Twelve tales by such masters as Chekhov, Tolstoy, Dostoevsky, Pushkin, others. Excellent word-for-word English translations on facing pages, plus teaching and study aids, Russian/English vocabulary, biographical/critical introductions, more. 416pp. 5⅜ × 8½. 26244-8 Pa. $7.95

PHILADELPHIA THEN AND NOW: 60 Sites Photographed in the Past and Present, Kenneth Finkel and Susan Oyama. Rare photographs of City Hall, Logan Square, Independence Hall, Betsy Ross House, other landmarks juxtaposed with contemporary views. Captures changing face of historic city. Introduction. Captions. 128pp. 8¼ × 11. 25790-8 Pa. $9.95

AIA ARCHITECTURAL GUIDE TO NASSAU AND SUFFOLK COUNTIES, LONG ISLAND, The American Institute of Architects, Long Island Chapter, and the Society for the Preservation of Long Island Antiquities. Comprehensive, well-researched and generously illustrated volume brings to life over three centuries of Long Island's great architectural heritage. More than 240 photographs with authoritative, extensively detailed captions. 176pp. 8¼ × 11. 26946-9 Pa. $14.95

NORTH AMERICAN INDIAN LIFE: Customs and Traditions of 23 Tribes, Elsie Clews Parsons (ed.). 27 fictionalized essays by noted anthropologists examine religion, customs, government, additional facets of life among the Winnebago, Crow, Zuni, Eskimo, other tribes. 480pp. 6⅛ × 9¼. 27377-6 Pa. $10.95

FRANK LLOYD WRIGHT'S HOLLYHOCK HOUSE, Donald Hoffmann. Lavishly illustrated, carefully documented study of one of Wright's most controversial residential designs. Over 120 photographs, floor plans, elevations, etc. Detailed perceptive text by noted Wright scholar. Index. 128pp. 9¼ × 10¾.
27133-1 Pa. $10.95

THE MALE AND FEMALE FIGURE IN MOTION: 60 Classic Photographic Sequences, Eadweard Muybridge. 60 true-action photographs of men and women walking, running, climbing, bending, turning, etc., reproduced from rare 19th-century masterpiece. vi + 121pp. 9 × 12. 24745-7 Pa. $10.95

1001 QUESTIONS ANSWERED ABOUT THE SEASHORE, N. J. Berrill and Jacquelyn Berrill. Queries answered about dolphins, sea snails, sponges, starfish, fishes, shore birds, many others. Covers appearance, breeding, growth, feeding, much more. 305pp. 5¼ × 8¼. 23366-9 Pa. $7.95

GUIDE TO OWL WATCHING IN NORTH AMERICA, Donald S. Heintzelman. Superb guide offers complete data and descriptions of 19 species: barn owl, screech owl, snowy owl, many more. Expert coverage of owl-watching equipment, conservation, migrations and invasions, etc. Guide to observing sites. 84 illustrations. xiii + 193pp. 5⅜ × 8½. 27344-X Pa. $7.95

MEDICINAL AND OTHER USES OF NORTH AMERICAN PLANTS: A Historical Survey with Special Reference to the Eastern Indian Tribes, Charlotte Erichsen-Brown. Chronological historical citations document 500 years of usage of plants, trees, shrubs native to eastern Canada, northeastern U.S. Also complete identifying information. 343 illustrations. 544pp. 6½ × 9¼. 25951-X Pa. $12.95

STORYBOOK MAZES, Dave Phillips. 23 stories and mazes on two-page spreads: Wizard of Oz, Treasure Island, Robin Hood, etc. Solutions. 64pp. 8¼ × 11.
23628-5 Pa. $2.95

NEGRO FOLK MUSIC, U.S.A., Harold Courlander. Noted folklorist's scholarly yet readable analysis of rich and varied musical tradition. Includes authentic versions of over 40 folk songs. Valuable bibliography and discography. xi + 324pp. 5⅜ × 8½. 27350-4 Pa. $7.95

MOVIE-STAR PORTRAITS OF THE FORTIES, John Kobal (ed.). 163 glamor, studio photos of 106 stars of the 1940s: Rita Hayworth, Ava Gardner, Marlon Brando, Clark Gable, many more. 176pp. 8⅜ × 11¼. 23546-7 Pa. $10.95

BENCHLEY LOST AND FOUND, Robert Benchley. Finest humor from early 30s, about pet peeves, child psychologists, post office and others. Mostly unavailable elsewhere. 73 illustrations by Peter Arno and others. 183pp. 5⅜ × 8½.
22410-4 Pa. $4.95

YEKL and THE IMPORTED BRIDEGROOM AND OTHER STORIES OF YIDDISH NEW YORK, Abraham Cahan. Film Hester Street based on Yekl (1896). Novel, other stories among first about Jewish immigrants on N.Y.'s East Side. 240pp. 5⅜ × 8½. 22427-9 Pa. $5.95

SELECTED POEMS, Walt Whitman. Generous sampling from *Leaves of Grass*. Twenty-four poems include "I Hear America Singing," "Song of the Open Road," "I Sing the Body Electric," "When Lilacs Last in the Dooryard Bloom'd," "O Captain! My Captain!"—all reprinted from an authoritative edition. Lists of titles and first lines. 128pp. 5³⁄₁₆ × 8¼. 26878-0 Pa. $1.00

THE BEST TALES OF HOFFMANN, E. T. A. Hoffmann. 10 of Hoffmann's most important stories: "Nutcracker and the King of Mice," "The Golden Flowerpot," etc. 458pp. 5⅜ × 8½. 21793-0 Pa. $8.95

FROM FETISH TO GOD IN ANCIENT EGYPT, E. A. Wallis Budge. Rich detailed survey of Egyptian conception of "God" and gods, magic, cult of animals, Osiris, more. Also, superb English translations of hymns and legends. 240 illustrations. 545pp. 5⅜ × 8½. 25803-3 Pa. $10.95

FRENCH STORIES/CONTES FRANÇAIS: A Dual-Language Book, Wallace Fowlie. Ten stories by French masters, Voltaire to Camus: "Micromegas" by Voltaire; "The Atheist's Mass" by Balzac; "Minuet" by de Maupassant; "The Guest" by Camus, six more. Excellent English translations on facing pages. Also French-English vocabulary list, exercises, more. 352pp. 5⅜ × 8½. 26443-2 Pa. $8.95

CHICAGO AT THE TURN OF THE CENTURY IN PHOTOGRAPHS: 122 Historic Views from the Collections of the Chicago Historical Society, Larry A. Viskochil. Rare large-format prints offer detailed views of City Hall, State Street, the Loop, Hull House, Union Station, many other landmarks, circa 1904–1913. Introduction. Captions. Maps. 144pp. 9⅜ × 12¼. 24656-6 Pa. $12.95

OLD BROOKLYN IN EARLY PHOTOGRAPHS, 1865–1929, William Lee Younger. Luna Park, Gravesend race track, construction of Grand Army Plaza, moving of Hotel Brighton, etc. 157 previously unpublished photographs. 165pp. 8⅞ × 11¾. 23587-4 Pa. $12.95

THE MYTHS OF THE NORTH AMERICAN INDIANS, Lewis Spence. Rich anthology of the myths and legends of the Algonquins, Iroquois, Pawnees and Sioux, prefaced by an extensive historical and ethnological commentary. 36 illustrations. 480pp. 5⅜ × 8½. 25967-6 Pa. $8.95

AN ENCYCLOPEDIA OF BATTLES: Accounts of Over 1,560 Battles from 1479 B.C. to the Present, David Eggenberger. Essential details of every major battle in recorded history from the first battle of Megiddo in 1479 B.C. to Grenada in 1984. List of Battle Maps. New Appendix covering the years 1967–1984. Index. 99 illustrations. 544pp. 6½ × 9¼. 24913-1 Pa. $14.95

SAILING ALONE AROUND THE WORLD, Captain Joshua Slocum. First man to sail around the world, alone, in small boat. One of great feats of seamanship told in delightful manner. 67 illustrations. 294pp. 5⅜ × 8½. 20326-3 Pa. $4.95

ANARCHISM AND OTHER ESSAYS, Emma Goldman. Powerful, penetrating, prophetic essays on direct action, role of minorities, prison reform, puritan hypocrisy, violence, etc. 271pp. 5⅜ × 8½. 22484-8 Pa. $5.95

MYTHS OF THE HINDUS AND BUDDHISTS, Ananda K. Coomaraswamy and Sister Nivedita. Great stories of the epics; deeds of Krishna, Shiva, taken from puranas, Vedas, folk tales; etc. 32 illustrations. 400pp. 5⅜ × 8½. 21759-0 Pa. $8.95

BEYOND PSYCHOLOGY, Otto Rank. Fear of death, desire of immortality, nature of sexuality, social organization, creativity, according to Rankian system. 291pp. 5⅜ × 8½. 20485-5 Pa. $7.95

A THEOLOGICO-POLITICAL TREATISE, Benedict Spinoza. Also contains unfinished Political Treatise. Great classic on religious liberty, theory of government on common consent. R. Elwes translation. Total of 421pp. 5⅜ × 8½. 20249-6 Pa. $7.95

MY BONDAGE AND MY FREEDOM, Frederick Douglass. Born a slave, Douglass became outspoken force in antislavery movement. The best of Douglass' autobiographies. Graphic description of slave life. 464pp. 5⅜ × 8½. 22457-0 Pa. $7.95

FOLLOWING THE EQUATOR: A Journey Around the World, Mark Twain. Fascinating humorous account of 1897 voyage to Hawaii, Australia, India, New Zealand, etc. Ironic, bemused reports on peoples, customs, climate, flora and fauna, politics, much more. 197 illustrations. 720pp. 5⅜ × 8½. 26113-1 Pa. $15.95

THE PEOPLE CALLED SHAKERS, Edward D. Andrews. Definitive study of Shakers: origins, beliefs, practices, dances, social organization, furniture and crafts, etc. 33 illustrations. 351pp. 5⅜ × 8½. 21081-2 Pa. $7.95

THE MYTHS OF GREECE AND ROME, H. A. Guerber. A classic of mythology, generously illustrated, long prized for its simple, graphic, accurate retelling of the principal myths of Greece and Rome, and for its commentary on their origins and significance. With 64 illustrations by Michelangelo, Raphael, Titian, Rubens, Canova, Bernini and others. 480pp. 5⅜ × 8½. 27584-1 Pa. $9.95

PSYCHOLOGY OF MUSIC, Carl E. Seashore. Classic work discusses music as a medium from psychological viewpoint. Clear treatment of physical acoustics, auditory apparatus, sound perception, development of musical skills, nature of musical feeling, host of other topics. 88 figures. 408pp. 5⅜ × 8½. 21851-1 Pa. $8.95

THE PHILOSOPHY OF HISTORY, Georg W. Hegel. Great classic of Western thought develops concept that history is not chance but rational process, the evolution of freedom. 457pp. 5⅜ × 8½. 20112-0 Pa. $8.95

THE BOOK OF TEA, Kakuzo Okakura. Minor classic of the Orient: entertaining, charming explanation, interpretation of traditional Japanese culture in terms of tea ceremony. 94pp. 5⅜ × 8½. 20070-1 Pa. $2.95

LIFE IN ANCIENT EGYPT, Adolf Erman. Fullest, most thorough, detailed older account with much not in more recent books, domestic life, religion, magic, medicine, commerce, much more. Many illustrations reproduce tomb paintings, carvings, hieroglyphs, etc. 597pp. 5⅜ × 8½. 22632-8 Pa. $9.95

SUNDIALS, Their Theory and Construction, Albert Waugh. Far and away the best, most thorough coverage of ideas, mathematics concerned, types, construction, adjusting anywhere. Simple, nontechnical treatment allows even children to build several of these dials. Over 100 illustrations. 230pp. 5⅜ × 8½. 22947-5 Pa. $5.95

DYNAMICS OF FLUIDS IN POROUS MEDIA, Jacob Bear. For advanced students of ground water hydrology, soil mechanics and physics, drainage and irrigation engineering, and more. 335 illustrations. Exercises, with answers. 784pp. 6⅛ × 9¼. 65675-6 Pa. $19.95

SONGS OF EXPERIENCE: Facsimile Reproduction with 26 Plates in Full Color, William Blake. 26 full-color plates from a rare 1826 edition. Includes "The Tyger," "London," "Holy Thursday," and other poems. Printed text of poems. 48pp. 5¼ × 7. 24636-1 Pa. $3.95

OLD-TIME VIGNETTES IN FULL COLOR, Carol Belanger Grafton (ed.). Over 390 charming, often sentimental illustrations, selected from archives of Victorian graphics—pretty women posing, children playing, food, flowers, kittens and puppies, smiling cherubs, birds and butterflies, much more. All copyright-free. 48pp. 9¼ × 12¼. 27269-9 Pa. $5.95

PERSPECTIVE FOR ARTISTS, Rex Vicat Cole. Depth, perspective of sky and sea, shadows, much more, not usually covered. 391 diagrams, 81 reproductions of drawings and paintings. 279pp. 5⅜ × 8½. 22487-2 Pa. $6.95

DRAWING THE LIVING FIGURE, Joseph Sheppard. Innovative approach to artistic anatomy focuses on specifics of surface anatomy, rather than muscles and bones. Over 170 drawings of live models in front, back and side views, and in widely varying poses. Accompanying diagrams. 177 illustrations. Introduction. Index. 144pp. 8⅜ × 11¼. 26723-7 Pa. $7.95

GOTHIC AND OLD ENGLISH ALPHABETS: 100 Complete Fonts, Dan X. Solo. Add power, elegance to posters, signs, other graphics with 100 stunning copyright-free alphabets: Blackstone, Dolbey, Germania, 97 more—including many lowercase, numerals, punctuation marks. 104pp. 8⅛ × 11. 24695-7 Pa. $6.95

HOW TO DO BEADWORK, Mary White. Fundamental book on craft from simple projects to five-bead chains and woven works. 106 illustrations. 142pp. 5⅜ × 8. 20697-1 Pa. $4.95

THE BOOK OF WOOD CARVING, Charles Marshall Sayers. Finest book for beginners discusses fundamentals and offers 34 designs. "Absolutely first rate . . . well thought out and well executed."—E. J. Tangerman. 118pp. 7¾ × 10⅝. 23654-4 Pa. $5.95

ILLUSTRATED CATALOG OF CIVIL WAR MILITARY GOODS: Union Army Weapons, Insignia, Uniform Accessories, and Other Equipment, Schuyler, Hartley, and Graham. Rare, profusely illustrated 1846 catalog includes Union Army uniform and dress regulations, arms and ammunition, coats, insignia, flags, swords, rifles, etc. 226 illustrations. 160pp. 9 × 12. 24939-5 Pa. $10.95

WOMEN'S FASHIONS OF THE EARLY 1900s: An Unabridged Republication of "New York Fashions, 1909," National Cloak & Suit Co. Rare catalog of mail-order fashions documents women's and children's clothing styles shortly after the turn of the century. Captions offer full descriptions, prices. Invaluable resource for fashion, costume historians. Approximately 725 illustrations. 128pp. 8⅜ × 11¼. 27276-1 Pa. $10.95

THE 1912 AND 1915 GUSTAV STICKLEY FURNITURE CATALOGS, Gustav Stickley. With over 200 detailed illustrations and descriptions, these two catalogs are essential reading and reference materials and identification guides for Stickley furniture. Captions cite materials, dimensions and prices. 112pp. 6½ × 9¼. 26676-1 Pa. $9.95

EARLY AMERICAN LOCOMOTIVES, John H. White, Jr. Finest locomotive engravings from early 19th century: historical (1804–74), main-line (after 1870), special, foreign, etc. 147 plates. 142pp. 11⅜ × 8¼. 22772-3 Pa. $8.95

THE TALL SHIPS OF TODAY IN PHOTOGRAPHS, Frank O. Braynard. Lavishly illustrated tribute to nearly 100 majestic contemporary sailing vessels: Amerigo Vespucci, Clearwater, Constitution, Eagle, Mayflower, Sea Cloud, Victory, many more. Authoritative captions provide statistics, background on each ship. 190 black-and-white photographs and illustrations. Introduction. 128pp. 8⅞ × 11¾. 27163-3 Pa. $12.95

EARLY NINETEENTH-CENTURY CRAFTS AND TRADES, Peter Stockham (ed.). Extremely rare 1807 volume describes to youngsters the crafts and trades of the day: brickmaker, weaver, dressmaker, bookbinder, ropemaker, saddler, many more. Quaint prose, charming illustrations for each craft. 20 black-and-white line illustrations. 192pp. 4⅜ × 6. 27293-1 Pa. $4.95

VICTORIAN FASHIONS AND COSTUMES FROM HARPER'S BAZAR, 1867–1898, Stella Blum (ed.). Day costumes, evening wear, sports clothes, shoes, hats, other accessories in over 1,000 detailed engravings. 320pp. 9⅜ × 12¼. 22990-4 Pa. $12.95

GUSTAV STICKLEY, THE CRAFTSMAN, Mary Ann Smith. Superb study surveys broad scope of Stickley's achievement, especially in architecture. Design philosophy, rise and fall of the Craftsman empire, descriptions and floor plans for many Craftsman houses, more. 86 black-and-white halftones. 31 line illustrations. Introduction. 208pp. 6½ × 9¼. 27210-9 Pa. $9.95

THE LONG ISLAND RAIL ROAD IN EARLY PHOTOGRAPHS, Ron Ziel. Over 220 rare photos, informative text document origin (1844) and development of rail service on Long Island. Vintage views of early trains, locomotives, stations, passengers, crews, much more. Captions. 8⅞ × 11¾. 26301-0 Pa. $13.95

THE BOOK OF OLD SHIPS: From Egyptian Galleys to Clipper Ships, Henry B. Culver. Superb, authoritative history of sailing vessels, with 80 magnificent line illustrations. Galley, bark, caravel, longship, whaler, many more. Detailed, informative text on each vessel by noted naval historian. Introduction. 256pp. 5⅜ × 8½. 27332-6 Pa. $6.95

TEN BOOKS ON ARCHITECTURE, Vitruvius. The most important book ever written on architecture. Early Roman aesthetics, technology, classical orders, site selection, all other aspects. Morgan translation. 331pp. 5⅜ × 8½. 20645-9 Pa. $8.95

THE HUMAN FIGURE IN MOTION, Eadweard Muybridge. More than 4,500 stopped-action photos, in action series, showing undraped men, women, children jumping, lying down, throwing, sitting, wrestling, carrying, etc. 390pp. 7⅞ × 10⅝. 20204-6 Clothbd. $24.95

TREES OF THE EASTERN AND CENTRAL UNITED STATES AND CANADA, William M. Harlow. Best one-volume guide to 140 trees. Full descriptions, woodlore, range, etc. Over 600 illustrations. Handy size. 288pp. 4½ × 6⅜. 20395-6 Pa. $4.95

SONGS OF WESTERN BIRDS, Dr. Donald J. Borror. Complete song and call repertoire of 60 western species, including flycatchers, juncoes, cactus wrens, many more—includes fully illustrated booklet. Cassette and manual 99913-0 $8.95

GROWING AND USING HERBS AND SPICES, Milo Miloradovich. Versatile handbook provides all the information needed for cultivation and use of all the herbs and spices available in North America. 4 illustrations. Index. Glossary. 236pp. 5⅜ × 8½. 25058-X Pa. $5.95

BIG BOOK OF MAZES AND LABYRINTHS, Walter Shepherd. 50 mazes and labyrinths in all—classical, solid, ripple, and more—in one great volume. Perfect inexpensive puzzler for clever youngsters. Full solutions. 112pp. 8⅛ × 11. 22951-3 Pa. $3.95

PIANO TUNING, J. Cree Fischer. Clearest, best book for beginner, amateur. Simple repairs, raising dropped notes, tuning by easy method of flattened fifths. No previous skills needed. 4 illustrations. 201pp. 5⅜ × 8½. 23267-0 Pa. $4.95

A SOURCE BOOK IN THEATRICAL HISTORY, A. M. Nagler. Contemporary observers on acting, directing, make-up, costuming, stage props, machinery, scene design, from Ancient Greece to Chekhov. 611pp. 5⅜ × 8½. 20515-0 Pa. $10.95

THE COMPLETE NONSENSE OF EDWARD LEAR, Edward Lear. All nonsense limericks, zany alphabets, Owl and Pussycat, songs, nonsense botany, etc., illustrated by Lear. Total of 320pp. 5⅜ × 8½. (USO) 20167-8 Pa. $5.95

VICTORIAN PARLOUR POETRY: An Annotated Anthology, Michael R. Turner. 117 gems by Longfellow, Tennyson, Browning, many lesser-known poets. "The Village Blacksmith," "Curfew Must Not Ring Tonight," "Only a Baby Small," dozens more, often difficult to find elsewhere. Index of poets, titles, first lines. xxiii + 325pp. 5⅜ × 8¼. 27044-0 Pa. $7.95

DUBLINERS, James Joyce. Fifteen stories offer vivid, tightly focused observations of the lives of Dublin's poorer classes. At least one, "The Dead," is considered a masterpiece. Reprinted complete and unabridged from standard edition. 160pp. 5³⁄₁₆ × 8¼. 26870-5 Pa. $1.00

THE HAUNTED MONASTERY and THE CHINESE MAZE MURDERS, Robert van Gulik. Two full novels by van Gulik, set in 7th-century China, continue adventures of Judge Dee and his companions. An evil Taoist monastery, seemingly supernatural events; overgrown topiary maze hides strange crimes. 27 illustrations. 328pp. 5⅜ × 8½. 23502-5 Pa. $7.95

THE BOOK OF THE SACRED MAGIC OF ABRAMELIN THE MAGE, translated by S. MacGregor Mathers. Medieval manuscript of ceremonial magic. Basic document in Aleister Crowley, Golden Dawn groups. 268pp. 5⅜ × 8½. 23211-5 Pa. $7.95

NEW RUSSIAN-ENGLISH AND ENGLISH-RUSSIAN DICTIONARY, M. A. O'Brien. This is a remarkably handy Russian dictionary, containing a surprising amount of information, including over 70,000 entries. 366pp. 4½ × 6⅛. 20208-9 Pa. $8.95

HISTORIC HOMES OF THE AMERICAN PRESIDENTS, Second, Revised Edition, Irvin Haas. A traveler's guide to American Presidential homes, most open to the public, depicting and describing homes occupied by every American President from George Washington to George Bush. With visiting hours, admission charges, travel routes. 175 photographs. Index. 160pp. 8¼ × 11. 26751-2 Pa. $10.95

NEW YORK IN THE FORTIES, Andreas Feininger. 162 brilliant photographs by the well-known photographer, formerly with *Life* magazine. Commuters, shoppers, Times Square at night, much else from city at its peak. Captions by John von Hartz. 181pp. 9¼ × 10¾. 23585-8 Pa. $12.95

INDIAN SIGN LANGUAGE, William Tomkins. Over 525 signs developed by Sioux and other tribes. Written instructions and diagrams. Also 290 pictographs. 111pp. 6⅛ × 9¼. 22029-X Pa. $3.50

ANATOMY: A Complete Guide for Artists, Joseph Sheppard. A master of figure drawing shows artists how to render human anatomy convincingly. Over 460 illustrations. 224pp. 8⅜ × 11¼. 27279-6 Pa. $9.95

MEDIEVAL CALLIGRAPHY: Its History and Technique, Marc Drogin. Spirited history, comprehensive instruction manual covers 13 styles (ca. 4th century thru 15th). Excellent photographs; directions for duplicating medieval techniques with modern tools. 224pp. 8⅜ × 11¼. 26142-5 Pa. $11.95

DRIED FLOWERS: How to Prepare Them, Sarah Whitlock and Martha Rankin. Complete instructions on how to use silica gel, meal and borax, perlite aggregate, sand and borax, glycerine and water to create attractive permanent flower arrangements. 12 illustrations. 32pp. 5⅜ × 8½. 21802-3 Pa. $1.00

EASY-TO-MAKE BIRD FEEDERS FOR WOODWORKERS, Scott D. Campbell. Detailed, simple-to-use guide for designing, constructing, caring for and using feeders. Text, illustrations for 12 classic and contemporary designs. 96pp. 5⅜ × 8½. 25847-5 Pa. $2.95

OLD-TIME CRAFTS AND TRADES, Peter Stockham. An 1807 book created to teach children about crafts and trades open to them as future careers. It describes in detailed, nontechnical terms 24 different occupations, among them coachmaker, gardener, hairdresser, lacemaker, shoemaker, wheelwright, copper-plate printer, milliner, trunkmaker, merchant and brewer. Finely detailed engravings illustrate each occupation. 192pp. 4⅝ × 6. 27398-9 Pa. $4.95

THE HISTORY OF UNDERCLOTHES, C. Willett Cunnington and Phyllis Cunnington. Fascinating, well-documented survey covering six centuries of English undergarments, enhanced with over 100 illustrations: 12th-century laced-up bodice, footed long drawers (1795), 19th-century bustles, 19th-century corsets for men, Victorian "bust improvers," much more. 272pp. 5⅜ × 8¼. 27124-2 Pa. $9.95

ARTS AND CRAFTS FURNITURE: The Complete Brooks Catalog of 1912, Brooks Manufacturing Co. Photos and detailed descriptions of more than 150 now very collectible furniture designs from the Arts and Crafts movement depict davenports, settees, buffets, desks, tables, chairs, bedsteads, dressers and more, all built of solid, quarter-sawed oak. Invaluable for students and enthusiasts of antiques, Americana and the decorative arts. 80pp. 6½ × 9¼. 27471-3 Pa. $7.95

HOW WE INVENTED THE AIRPLANE: An Illustrated History, Orville Wright. Fascinating firsthand account covers early experiments, construction of planes and motors, first flights, much more. Introduction and commentary by Fred C. Kelly. 76 photographs. 96pp. 8¼ × 11. 25662-6 Pa. $7.95

THE ARTS OF THE SAILOR: Knotting, Splicing and Ropework, Hervey Garrett Smith. Indispensable shipboard reference covers tools, basic knots and useful hitches; handsewing and canvas work, more. Over 100 illustrations. Delightful reading for sea lovers. 256pp. 5⅜ × 8½. 26440-8 Pa. $6.95

FRANK LLOYD WRIGHT'S FALLINGWATER: The House and Its History, Second, Revised Edition, Donald Hoffmann. A total revision—both in text and illustrations—of the standard document on Fallingwater, the boldest, most personal architectural statement of Wright's mature years, updated with valuable new material from the recently opened Frank Lloyd Wright Archives. "Fascinating"—*The New York Times*. 116 illustrations. 128pp. 9¼ × 10¾. 27430-6 Pa. $10.95

PHOTOGRAPHIC SKETCHBOOK OF THE CIVIL WAR, Alexander Gardner. 100 photos taken on field during the Civil War. Famous shots of Manassas, Harper's Ferry, Lincoln, Richmond, slave pens, etc. 244pp. 10⅝ × 8¼.
22731-6 Pa. $9.95

FIVE ACRES AND INDEPENDENCE, Maurice G. Kains. Great back-to-the-land classic explains basics of self-sufficient farming. The one book to get. 95 illustrations. 397pp. 5⅜ × 8½. 20974-1 Pa. $6.95

SONGS OF EASTERN BIRDS, Dr. Donald J. Borror. Songs and calls of 60 species most common to eastern U.S.: warblers, woodpeckers, flycatchers, thrushes, larks, many more in high-quality recording. Cassette and manual 99912-2 $8.95

A MODERN HERBAL, Margaret Grieve. Much the fullest, most exact, most useful compilation of herbal material. Gigantic alphabetical encyclopedia, from aconite to zedoary, gives botanical information, medical properties, folklore, economic uses, much else. Indispensable to serious reader. 161 illustrations. 888pp. 6½ × 9¼. 2-vol. set. (USO) Vol. I: 22798-7 Pa. $9.95
Vol. II: 22799-5 Pa. $9.95

HIDDEN TREASURE MAZE BOOK, Dave Phillips. Solve 34 challenging mazes accompanied by heroic tales of adventure. Evil dragons, people-eating plants, bloodthirsty giants, many more dangerous adversaries lurk at every twist and turn. 34 mazes, stories, solutions. 48pp. 8¼ × 11. 24566-7 Pa. $2.95

LETTERS OF W. A. MOZART, Wolfgang A. Mozart. Remarkable letters show bawdy wit, humor, imagination, musical insights, contemporary musical world; includes some letters from Leopold Mozart. 276pp. 5⅜ × 8½. 22859-2 Pa. $6.95

BASIC PRINCIPLES OF CLASSICAL BALLET, Agrippina Vaganova. Great Russian theoretician, teacher explains methods for teaching classical ballet. 118 illustrations. 175pp. 5⅜ × 8½. 22036-2 Pa. $3.95

THE JUMPING FROG, Mark Twain. Revenge edition. The original story of The Celebrated Jumping Frog of Calaveras County, a hapless French translation, and Twain's hilarious "retranslation" from the French. 12 illustrations. 66pp. 5⅜ × 8½.
22686-7 Pa. $3.50

BEST REMEMBERED POEMS, Martin Gardner (ed.). The 126 poems in this superb collection of 19th- and 20th-century British and American verse range from Shelley's "To a Skylark" to the impassioned "Renascence" of Edna St. Vincent Millay and to Edward Lear's whimsical "The Owl and the Pussycat." 224pp. 5⅜ × 8½.
27165-X Pa. $3.95

COMPLETE SONNETS, William Shakespeare. Over 150 exquisite poems deal with love, friendship, the tyranny of time, beauty's evanescence, death and other themes in language of remarkable power, precision and beauty. Glossary of archaic terms. 80pp. 5³⁄₁₆ × 8¼. 26686-9 Pa. $1.00

BODIES IN A BOOKSHOP, R. T. Campbell. Challenging mystery of blackmail and murder with ingenious plot and superbly drawn characters. In the best tradition of British suspense fiction. 192pp. 5⅜ × 8½. 24720-1 Pa. $5.95

THE WIT AND HUMOR OF OSCAR WILDE, Alvin Redman (ed.). More than 1,000 ripostes, paradoxes, wisecracks: Work is the curse of the drinking classes; I can resist everything except temptation; etc. 258pp. 5⅜ × 8½. 20602-5 Pa. $4.95

SHAKESPEARE LEXICON AND QUOTATION DICTIONARY, Alexander Schmidt. Full definitions, locations, shades of meaning in every word in plays and poems. More than 50,000 exact quotations. 1,485pp. 6½ × 9¼. 2-vol. set.
Vol. 1: 22726-X Pa. $15.95
Vol. 2: 22727-8 Pa. $15.95

SELECTED POEMS, Emily Dickinson. Over 100 best-known, best-loved poems by one of America's foremost poets, reprinted from authoritative early editions. No comparable edition at this price. Index of first lines. 64pp. 5³⁄₁₆ × 8¼.
26466-1 Pa. $1.00

CELEBRATED CASES OF JUDGE DEE (DEE GOONG AN), translated by Robert van Gulik. Authentic 18th-century Chinese detective novel; Dee and associates solve three interlocked cases. Led to van Gulik's own stories with same characters. Extensive introduction. 9 illustrations. 237pp. 5⅜ × 8½.
23337-5 Pa. $5.95

THE MALLEUS MALEFICARUM OF KRAMER AND SPRENGER, translated by Montague Summers. Full text of most important witchhunter's "bible," used by both Catholics and Protestants. 278pp. 6⅝ × 10. 22802-9 Pa. $10.95

SPANISH STORIES/CUENTOS ESPAÑOLES: A Dual-Language Book, Angel Flores (ed.). Unique format offers 13 great stories in Spanish by Cervantes, Borges, others. Faithful English translations on facing pages. 352pp. 5⅜ × 8½.
25399-6 Pa. $7.95

THE CHICAGO WORLD'S FAIR OF 1893: A Photographic Record, Stanley Appelbaum (ed.). 128 rare photos show 200 buildings, Beaux-Arts architecture, Midway, original Ferris Wheel, Edison's kinetoscope, more. Architectural emphasis; full text. 116pp. 8¼ × 11. 23990-X Pa. $9.95

OLD QUEENS, N.Y., IN EARLY PHOTOGRAPHS, Vincent F. Seyfried and William Asadorian. Over 160 rare photographs of Maspeth, Jamaica, Jackson Heights, and other areas. Vintage views of DeWitt Clinton mansion, 1939 World's Fair and more. Captions. 192pp. 8⅞ × 11. 26358-4 Pa. $12.95

CAPTURED BY THE INDIANS: 15 Firsthand Accounts, 1750–1870, Frederick Drimmer. Astounding true historical accounts of grisly torture, bloody conflicts, relentless pursuits, miraculous escapes and more, by people who lived to tell the tale. 384pp. 5⅜ × 8½. 24901-8 Pa. $7.95

THE WORLD'S GREAT SPEECHES, Lewis Copeland and Lawrence W. Lamm (eds.). Vast collection of 278 speeches of Greeks to 1970. Powerful and effective models; unique look at history. 842pp. 5⅜ × 8½. 20468-5 Pa. $12.95

THE BOOK OF THE SWORD, Sir Richard F. Burton. Great Victorian scholar/adventurer's eloquent, erudite history of the "queen of weapons"—from prehistory to early Roman Empire. Evolution and development of early swords, variations (sabre, broadsword, cutlass, scimitar, etc.), much more. 336pp. 6⅛ × 9¼. 25434-8 Pa. $8.95

CATALOG OF DOVER BOOKS

AUTOBIOGRAPHY: The Story of My Experiments with Truth, Mohandas K. Gandhi. Boyhood, legal studies, purification, the growth of the Satyagraha (nonviolent protest) movement. Critical, inspiring work of the man responsible for the freedom of India. 480pp. 5⅜ × 8½. (USO) 24593-4 Pa. $6.95

CELTIC MYTHS AND LEGENDS, T. W. Rolleston. Masterful retelling of Irish and Welsh stories and tales. Cuchulain, King Arthur, Deirdre, the Grail, many more. First paperback edition. 58 full-page illustrations. 512pp. 5⅜ × 8½. 26507-2 Pa. $9.95

THE PRINCIPLES OF PSYCHOLOGY, William James. Famous long course complete, unabridged. Stream of thought, time perception, memory, experimental methods; great work decades ahead of its time. 94 figures. 1,391pp. 5⅜ × 8½. 2-vol. set.
Vol. I: 20381-6 Pa. $12.95
Vol. II: 20382-4 Pa. $12.95

THE WORLD AS WILL AND REPRESENTATION, Arthur Schopenhauer. Definitive English translation of Schopenhauer's life work, correcting more than 1,000 errors, omissions in earlier translations. Translated by E. F. J. Payne. Total of 1,269pp. 5⅜ × 8½. 2-vol. set. Vol. 1: 21761-2 Pa. $10.95
Vol. 2: 21762-0 Pa. $11.95

MAGIC AND MYSTERY IN TIBET, Madame Alexandra David-Neel. Experiences among lamas, magicians, sages, sorcerers, Bonpa wizards. A true psychic discovery. 32 illustrations. 321pp. 5⅜ × 8½. (USO) 22682-4 Pa. $7.95

THE EGYPTIAN BOOK OF THE DEAD, E. A. Wallis Budge. Complete reproduction of Ani's papyrus, finest ever found. Full hieroglyphic text, interlinear transliteration, word-for-word translation, smooth translation. 533pp. 6½ × 9¼. 21866-X Pa. $9.95

MATHEMATICS FOR THE NONMATHEMATICIAN, Morris Kline. Detailed, college-level treatment of mathematics in cultural and historical context, with numerous exercises. Recommended Reading Lists. Tables. Numerous figures. 641pp. 5⅜ × 8½. 24823-2 Pa. $11.95

THEORY OF WING SECTIONS: Including a Summary of Airfoil Data, Ira H. Abbott and A. E. von Doenhoff. Concise compilation of subsonic aerodynamic characteristics of NACA wing sections, plus description of theory. 350pp. of tables. 693pp. 5⅜ × 8½. 60586-8 Pa. $13.95

THE RIME OF THE ANCIENT MARINER, Gustave Doré, S. T. Coleridge. Doré's finest work; 34 plates capture moods, subtleties of poem. Flawless full-size reproductions printed on facing pages with authoritative text of poem. "Beautiful. Simply beautiful."—*Publisher's Weekly*. 77pp. 9¼ × 12. 22305-1 Pa. $5.95

NORTH AMERICAN INDIAN DESIGNS FOR ARTISTS AND CRAFTSPEOPLE, Eva Wilson. Over 360 authentic copyright-free designs adapted from Navajo blankets, Hopi pottery, Sioux buffalo hides, more. Geometrics, symbolic figures, plant and animal motifs, etc. 128pp. 8⅜ × 11. (EUK) 25341-4 Pa. $6.95

SCULPTURE: Principles and Practice, Louis Slobodkin. Step-by-step approach to clay, plaster, metals, stone; classical and modern. 253 drawings, photos. 255pp. 8⅛ × 11. 22960-2 Pa. $9.95

THE INFLUENCE OF SEA POWER UPON HISTORY, 1660-1783, A. T. Mahan. Influential classic of naval history and tactics still used as text in war colleges. First paperback edition. 4 maps. 24 battle plans. 640pp. 5⅜ × 8½.
25509-3 Pa. $12.95

THE STORY OF THE TITANIC AS TOLD BY ITS SURVIVORS, Jack Winocour (ed.). What it was really like. Panic, despair, shocking inefficiency, and a little heroism. More thrilling than any fictional account. 26 illustrations. 320pp. 5⅜ × 8½. 20610-6 Pa. $7.95

FAIRY AND FOLK TALES OF THE IRISH PEASANTRY, William Butler Yeats (ed.). Treasury of 64 tales from the twilight world of Celtic myth and legend: "The Soul Cages," "The Kildare Pooka," "King O'Toole and his Goose," many more. Introduction and Notes by W. B. Yeats. 352pp. 5⅜ × 8½. 26941-8 Pa. $7.95

BUDDHIST MAHAYANA TEXTS, E. B. Cowell and Others (eds.). Superb, accurate translations of basic documents in Mahayana Buddhism, highly important in history of religions. The Buddha-karita of Asvaghosha, Larger Sukhavativyuha, more. 448pp. 5⅜ × 8½. 25552-2 Pa. $9.95

ONE TWO THREE . . . INFINITY: Facts and Speculations of Science, George Gamow. Great physicist's fascinating, readable overview of contemporary science: number theory, relativity, fourth dimension, entropy, genes, atomic structure, much more. 128 illustrations. Index. 352pp. 5⅜ × 8½. 25664-2 Pa. $7.95

ENGINEERING IN HISTORY, Richard Shelton Kirby, et al. Broad, nontechnical survey of history's major technological advances: birth of Greek science, industrial revolution, electricity and applied science, 20th-century automation, much more. 181 illustrations. ". . . excellent . . ."—Isis. Bibliography. vii + 530pp. 5⅜ × 8¼.
26412-2 Pa. $13.95

Prices subject to change without notice.

Available at your book dealer or write for free catalog to Dept. GI, Dover Publications, Inc., 31 East 2nd St., Mineola, N.Y. 11501. Dover publishes more than 500 books each year on science, elementary and advanced mathematics, biology, music, art, literary history, social sciences and other areas.